Max Planck

The Universe in the Light of Modern Physics

Edited by Vesselin Petkov

MINKOWSKI
Institute Press

Max Karl Ernst Ludwig Planck
23 April 1858 – 4 October 1947

Cover: https://commons.wikimedia.org/wiki/File:Max_Planck_
1933.jpg#/media/File:Max_Planck_1933.jpg

ISBN: 978-1-927763-92-6 (softcover)
ISBN: 978-1-927763-93-3 (ebook)

Minkowski Institute Press
Montreal, Quebec, Canada
http://minkowskiinstitute.org/mip/

For information on all Minkowski Institute Press publications
visit our website at http://minkowskiinstitute.org/mip/books/

PREFACE

This volume includes new publications of Max Planck's book *The Universe in the Light of Modern Physics*,[1] and four papers:[2]

- Phantom Problems in Science

- The Meaning and Limits of Exact Science

- The Concept of Causality in Physics

- Religion and Natural Science

In the book *The Universe in the Light of Modern Physics*, footnotes containing the abbreviation [TRANS.] are notes by the translator.

My initial idea to add comments to reflect advancements in modern physics since the time Planck wrote his book *The Universe in the Light of Modern Physics* was abandoned for two reasons – (i) comments (footnotes) especially on advancements in cosmology would be unreasonably long; (ii) the discussions of most of the issues in the book are still valid and valuable today. I added only three footnotes – EDITOR'S NOTES.

---

[1]Originally published: Max Planck, *The Universe in the Light of Modern Physics*. Translated by W. H. Johnston (George Allen & Unwin Ltd, London 1931).

[2]Originally published in Max Planck, *Scientific Autobiography And Other Papers*. Translated by F. Gaynor (Williams & Norgate Ltd, London 1950).

February 9, 2020

Vesselin Petkov
Minkowski Institute

CONTENTS

iii

iv

THE UNIVERSE IN THE LIGHT OF MODERN PHYSICS

§ 1

Physics is an exact Science and hence depends upon measurement, while all measurement itself requires sense-perception. Consequently all the ideas employed in Physics are derived from the world of sense-perception. It follows from this that the laws of Physics ultimately refer to events in the world of the senses; and in view of this fact many scientists and philosophers tend to the belief that at bottom Physics is concerned exclusively with this particular world. What they have in mind, of course, is the world of man's senses. On this view, for example, what is called an "Object" in ordinary parlance is, when regarded from the standpoint of Physics, simply a combination of different sense-data localized in one place. It is worth pointing out that this view cannot be refuted by logic, since logic itself is unable to lead us beyond the confines of our own senses; it cannot even compel one to admit the independent existence of others outside oneself.

In Physics, however, as in every other science, common sense alone is not supreme; there must also be a place for Reason. Further, the mere absence of logical contradiction does not necessarily imply that everything is reasonable. Now reason tells us that if we turn our back upon a so-called object and cease to attend to it, the object still continues to ex-

2

ist. Reason tells us further that both the individual man and mankind as a whole, together with the entire world which we apprehend through our senses, is no more than a tiny fragment in the vastness of Nature, whose laws are in no way affected by any human brain. On the contrary, they existed long before there was any life on earth, and will continue to exist long after the last physicist has perished.

It is considerations of this kind, and not any logical argument, that compel us to assume the existence of another world of reality behind the world of the senses; a world which has existence independent of man, and which can only be perceived indirectly through the medium of the world of the senses, and by means of certain symbols which our senses allow us to apprehend. It is as though we were compelled to contemplate a certain object in which we are interested through spectacles of whose optical properties we were entirely ignorant.

If the reader experiences difficulty in following this argument, and finds himself unable to accept the idea of a real world which at the same time is expressly asserted to lie beyond our senses, we might point out that there is a vast difference between a physical theory complete in every detail, and the construction of such a theory. In the former case the content of the theory can be analysed exactly, so that it is possible to prove at every point that the notions which we apply to the world of sense are adequate to the formulation of this theory; in the latter case we must develop a theory from a number of individual measurements. The second problem is very much more difficult, while the history of Physics shows that whenever it has been solved, this has been done on the assumption of a real world independent of our senses; and it seems reasonably certain that this will continue to be the case in the future.

But besides the world of sense and the real world, there is also a third world which must be carefully distinguished from these: this is the world of Physics. It differs from the two others because it is a deliberate hypothesis put forward by a finite human mind; and as such, it is subject to change and to a kind of evolution. Thus the function of this world of Physics may be described in two ways, according as it is related to the real world, or to the world of the senses. In the first case the problem is to apprehend the real world as completely as possible; in the second, to describe the world of the senses in the simplest possible terms. There is no need, however, to assign superior merit to either of these formulations, since each of them, taken by itself alone, is incomplete and unsatisfactory. On the one hand, the real world cannot be apprehended directly at all; while on the other no definite answer is possible to the question: Which is the simplest description of a given number of interdependent sense-perceptions? In the history of Physics it has happened more than once that, of two descriptions, one was for a time considered the more complicated but was later discovered to be the simpler of the two.

The essential point therefore is that these two formulations of the problem, when practically applied, shall be complementary to each other and not contradictory. The first is an indispensable aid to the groping imagination of the investigator, supplying him with ideas without which his work remains unfruitful; the second provides him with a firm foundation of facts. In actual practice individual physicists are influenced in their investigations by their personal preference for metaphysical, or for positivist, ideas. But besides the metaphysicians and the positivists there is a third group of students who investigate the world from the physical point of view. They differ from the first two groups in being interested not so much in the relation between the world of physics on the one hand, and the real world and the world of sense-data on the other, as in the internal consistency and logical structure of the world of physics. These men form the axiomatic school, whose activity is as necessary and useful as is that of the others. At the same time, they are equally exposed to the danger of specialization which, in their case, would lead to a barren formalism taking the place of a fuller understanding of the world of Physics. For as soon as contact with reality has been lost, physical law ceases to be felt as the relation between a number of magnitudes which have been ascertained independently of one another, and becomes a mere definition by which one of these magnitudes is derived from the others. In this method there is a particular attraction, due to the fact that a physical magnitude can be defined far more exactly by means of an equation than by means of measurement. But at the same time, this method amounts to a renunciation of the true meaning of magnitude; while it must also be remembered that confusion and misunderstanding result when the same name is retained in order to denote a changed meaning.

We see, then, how physicists are at work in different directions and from different standpoints in elaborating a systematic view of the world of Physics. Nevertheless the aim of all these endeavours is the same, and consists in establishing a law which connects the events of the world of sense with one another and with those of the real world. Naturally, these different tendencies predominated in turn at different stages of history. Whenever the physical world presented a stable appearance, as in the second half of the last century, the metaphysical view tended to predominate, and it was believed that a complete grasp of the real world was relatively near. Conversely, in times of change and insecurity like the present, positivism tends to occupy the foreground; for in such times a careful student will tend to seek support where he can find real security; and this is to be found precisely in the events of the world of the senses.

Now if we consider the different forms which the view of the physical world has taken in the course of history, and if we look for the peculiarities which characterized these changes, two facts will strike us with special force. First, it is plain that when regarded as a whole, all the changes in the different views of the world of Physics do not constitute a rhythmical swing of the pendulum. On the contrary, we find a clear course of evolution making more or less steady progress in a definite direction; progress which is best described by saying that it adds to the content of the world of sense, rendering our knowledge more profound and giving us a firmer grasp of it. The most striking instance of this is found in the practical application of Physics. Not even the most confirmed sceptic can deny that we see and hear at a greater distance and command greater forces and speeds than an earlier generation; while it is equally certain that this progress is an enduring increase of knowledge, which is in no danger of being described as an error and rejected at any future date.

Secondly, it is a very striking fact that the impulse towards simplification and improvement of the world-picture of Physics was due in each instance to some kind of novel observation that is, to some event in the world of sense. But at the same moment the structure of this physical world consistently moved farther and farther away from the world of sense and lost its former anthropomorphic character. Still further, physical sensations have been progressively eliminated, as for example in physical optics, in which the human eye no longer plays any part at all. Thus the physical world has become progressively more and more abstract; purely formal mathematical operations play a growing part, while qualitative differences tend to be explained more and more by means of quantitative differences.

Now we have already pointed out that the physical view of the world has been continually perfected and also related to the world of sense. If this fact is added to those mentioned in the last paragraph, the result is extraordinarily striking; at first, indeed, it appears completely paradoxical. Of this apparent paradox there is, in my opinion, only one rational explanation. This consists in saying that as the view of the physical world is perfected, it simultaneously recedes from the world of sense; and this process is tantamount to an approach to the world of reality. I have no logical proof on which to base this opinion; it is impossible to demonstrate the existence of the real world by purely rational methods: but at the same time it is equally impossible ever to refute it by logical methods. The final decision must rest upon a common-sense view of the world, and the old maxim still remains true that that world-view is the best which is the most fruitful. Physics would occupy an exceptional position among all the other sciences if it did not recognize the rule that the most far-reaching and valuable results of investigation can only be obtained by following a road leading to a goal which is theoretically unobtainable. This goal is the apprehension of true reality.

§ 2

What changes have taken place in the physical view of the world during the last twenty years? We all know that the changes which have occurred during this period are among the most profound that have ever arisen in the evolution of any science; we also know that the process of change has not yet come to an end. Nevertheless it would appear that in this flux of change certain characteristic forms of the structure of this new world are beginning to crystallize; and it is certainly worth while to attempt a description of these forms, if only in order to suggest certain improvements.

If we compare the old theory with the new, we find that the process of tracing back all qualitative distinctions to quantitative distinctions has been advanced very considerably. All the various chemical phenomena, for example, have now been explained by numerical and spatial relations. According to the modern view there are no more than two ultimate substances, namely positive and negative electricity. Each of these consists of a number of minute particles, similar in nature and with similar charges of an opposite character; the positive particle is called the proton, the negative the electron. Every chemical atom that is electrically neutral consists of a number of protons cohering with one another, and of a similar number of electrons, some of which are firmly fixed to the protons, together with which they form the nucleus of the atom, while the rest revolve around the nucleus.

Thus the Hydrogen atom, the smallest of all, has one proton for nucleus and one electron revolving round the nucleus; while the largest atom, Uranium, contains 238 protons and 238 electrons; but only 92 electrons revolve round the nucleus while the others are fixed in it. Between these two atoms lie all the other elements, with many kinds of different combinations. The chemical properties of an element depend, not on the total number of its protons or electrons, but on the number of revolving electrons, which yield the atomic number of the element.

Apart from this important advance, which is however merely the successful application of an idea first evolved many centuries ago, there are two completely new ideas which distinguish the modern conception of the world from its predecessor; these are the Theory of Relativity, and the Quantum Theory. It is these two ideas which are peculiarly characteristic of the new world of Physics. The fact that they appeared in science almost simultaneously is something of a coincidence; for their content, as well as their practical effect upon the structure of the physical view of the world, are entirely different.

The Theory of Relativity seemed at first to introduce a certain amount of confusion into the traditional ideas of Time and Space; in the long run, however, it has proved to be the completion and culmination of the structure of classical Physics. To express the positive result of the Special Theory of Relativity in a single word, it might be described as the fusion of Time and Space in one unitary concept. It is not, of course, asserted that Time and Space are absolutely similar in nature; their relation resembles that between a real number and an imaginary number, when these are combined together to form the unified concept of a complex number. Looked at in this way, Einstein's work for Physics closely resembles that of Gauss for Mathematics.[1] We might further continue the comparison by saying that the transition from the Special to the General Theory of Relativity is the counterpart in Physics to the transition from linear functions to the general theory of functions in mathematics.

Few comparisons are entirely exact, and the present is no exception to the rule. At the same time it gives a good idea of the fact that the introduction of the Theory of Relativity into the physical view of the world is one of the most

---

[1]EDITOR'S NOTE: Here Planck is not precise – not Einstein's but Minkowski's "work for Physics closely resembles that of Gauss for Mathematics." It is puzzling why Planck credited Einstein for something which Minkowski did, given that Planck had first-hand information of Minkowski's publications where he demonstrated that the relation between space and time in spacetime (Minkowski called it die *Welt* – the World) "resembles that between a real number and an imaginary number." Moreover, Planck had very likely been aware of Einstein's initial negative reaction to the "fusion of Time and Space in one unitary concept," which is Minkowski's achievement – "Since the mathematicians have invaded the relativity theory, I do not understand it myself any more" (quoted from: A. Sommerfeld, "To Albert Einstein's Seventieth Birthday." In: *Albert Einstein: Philosopher-Scientist*. P. A. Schilpp, ed., 3rd ed. (Open Court, Illinois 1969) pp. 99-105, p. 102). Also, two things (see http://www.minkowskiinstitute.org/born.html) appear to indicate that Minkowski arrived independently at what Einstein called special relativity and at the notion of spacetime, but Einstein and Poincaré published first. Minkowski did not publish his results earlier since he was developing the four-dimensional formalism of spacetime physics reported in 1907 and published in 1908 as a 59-page treatise (it is this formalism that we now use). As Max Born recalled, Minkowski "did not publish them because he wished first to work out the mathematical structure in all its splendour" (see the above link and the references therein).

important steps towards conferring unity and completeness. This appears clearly in the results of the Theory of Relativity, especially in the fusing of momentum and energy, in the identification of the concept of mass with the concept of energy, of inertial with ponderable mass, and in the reduction of the laws of gravitation to Riemann's geometry.

Brief though these main outlines are, they contain a vast mass of new knowledge. The new ideas mentioned apply to all natural events great and small, beginning with radioactive atoms emanating waves and corpuscles, and ending with the movements of celestial bodies millions of light-years away.

The last word on the Theory of Relativity probably still remains to be said. Surprises may yet await us, especially when we consider that the problem of amalgamating Electrodynamics with Mechanics has not yet been definitely solved. Again, the cosmological implications of the Theory of Relativity have not yet been fully cleared up, the chief reason being that everything depends upon the question whether or not the matter of outer space possesses a finite density; this question has not yet been answered. But whatever reply is eventually given to these questions, nothing will alter the fact that the Principle of Relativity has advanced the classical physical theory to its highest stage of completion, and that its world-view is rounded off in a very satisfactory manner.

This fact will perhaps be a sufficient reason for devoting no more time to the Theory of Relativity; I might also point out that there are many treatises on the Theory adapted to the requirements of readers of every kind.

§ 3

The idea of the universe as thus far described appeared almost perfectly adapted to its purpose; but this state of affairs has suddenly been upset by the Quantum Theory. Here again I shall attempt to describe the characteristic idea of this hypothesis in one word. We may say, then, that its essence consists in the fact that it introduces a new and universal constant, namely the elementary Quantum of Action. It was this constant which, like a new and mysterious messenger from the real world, insisted on turning up in every kind of measurement, and continued to claim a place for itself. On the other hand, it seemed so incompatible with the traditional view of the universe provided by Physics that it eventually destroyed the framework of this older view.

For a time it seemed that a complete collapse of classical Physics was not beyond the bounds of possibility; gradually however it appeared, as had been confidently expected by all who believed in the steady advance of science, that the introduction of the Quantum Theory led not to the destruction of Physics, but to a somewhat profound reconstruction, in the course of which the whole science was rendered more universal. For if the Quantum of Action is assumed to be infinitely small, Quantum Physics becomes merged in classical Physics. In fact the foundations of the structure of classical Physics not only proved unshakable, but actually were rendered firmer through the incorporation of the new ideas. The best course, therefore, will be first to examine the latter.

It will be best to begin by enumerating the essential component features. These are the universal constants, e.g. the gravitational constant, the velocity of light, the mass and charge of electrons and protons. These are perhaps the most tangible symbols of a real world, and they retain their meaning unchanged in the new view of the universe. Further, we may mention the great principles of the conservation of energy and of momentum, which, although they were under suspicion for a time, have eventually emerged unimpaired. It should be emphasized that in this process of transition these principles were proved to be something more than mere definitions, as some members of the Axiomatic School would like to believe. Further, we may mention the main laws of thermodynamics, and especially the second law, which through the introduction of an absolute value for entropy obtained a more exact formulation than it possessed in classical Physics. Lastly we may point to the Principle of Relativity, which has proved itself a reliable and eloquent guide in the new regions of Quantum Physics.

The question may now be asked whether modern Physics differs at all from the older Physics, if all these foundations of classical Physics have remained untouched. It is easy to find an answer to this question by examining the elementary Quantum of Action somewhat more closely. It implies that in principle an equation can be established between energy and frequency; $E = h\nu$.[2] It is this equation which classical Physics utterly fails to explain. The fact itself is so baffling because energy and frequency possess different dimensions; energy is a dynamic magnitude, whereas frequency is a kinematic magnitude. This fact in itself, however, does not contain a contradiction. The Quantum Theory postulates a direct connection between dynamics and kinematics; this connection is due to the fact that the unit of energy, and consequently the unit of mass, are based upon the units of length and of time; thus the connection, so far from being a contradiction, enriches and rounds off the classical theory. There is, nevertheless, a direct contradiction, which renders the new theory incompatible with the classical theory. The following considerations make clear this contradiction. Frequency is a local magnitude, and has a definite meaning only for a certain point in space; this is true alike of mechanical, electric and magnetic vibrations, so that all that is requisite is to observe the point

---

[2]In this equation E stands for Energy, and ν for Frequency, that is the number of vibrations per second. For example, light vibrations range from about 400 million million per second to about 800 million million. h represents "Planck's Constant," discovered by the author of this work. It is an unchanging or invariable quantity, and extremely minute, its value being 655 preceded by 26 decimal places. [TRANS.]

in question for a sufficient time. Energy on the other hand is
an additive quantity; so that according to the classical theory
it is meaningless to speak of energy at a certain point, since
it is essential to state the physical system the energy of which
is tinder discussion; just as it is similarly impossible to speak
of a definite velocity unless the system be indicated to which
velocity is referred. Now we are at liberty to choose whatever
physical system we please, either little or great; and consequently the value of the energy is always to a certain extent
arbitrary. The difficulty, then, consists in the fact that this
arbitrary energy is supposed to be equated with a localized
frequency. The gulf between these two concepts should now
be clearly apparent: and in order to bridge this gulf a step
of fundamental importance must be taken. This step does
imply a break with those assumptions which classical Physics
has always regarded and employed as axiomatic.

Hitherto it had been believed that the only kind of causality with which any system of Physics could operate was one
in which all the events of the physical world – by which, as
usual, I mean not the real world but the world-view of Physics
– might be explained as being composed of local events taking
place in a number of individual and infinitely small parts of
Space. It was further believed that each of these elementary
events was completely determined by a set of laws without respect to the other events; and was determined exclusively by
the local events in its immediate temporal and spatial vicin-
ity. Let us take a concrete instance of sufficiently general
application. We will assume that the physical system under
consideration consists of a system of particles, moving in a
conservative field of force of constant total energy. Then according to classical Physics each individual particle at any
time is in a definite state; that is, it has a definite position
and a definite velocity, and its movement can be calculated
with perfect exactness from its initial state and from the local
properties of the field of force in those parts of Space through

which the particle passes in the course of its movement. If these data are known, we need know nothing else about the remaining properties of the system of particles under consideration.

In modern mechanics matters are wholly different. According to modern mechanics, merely local relations are no more sufficient for the formulation of the law of motion than would be the microscopic investigation of the different parts of a picture in order to make clear its meaning. On the contrary, it is impossible to obtain an adequate version of the laws for which we are looking, unless the physical system is regarded *as a Whole*. According to modern mechanics, each individual particle of the system, in a certain sense, at any one time, exists simultaneously in every part of the space occupied by the system. This simultaneous existence applies not merely to the field of force with which it is surrounded, but also to its mass and its charge.

Thus we see that nothing less is at stake here than the concept of the particle the most elementary concept of classical mechanics. We are compelled to give up the earlier essential meaning of this idea; only in a number of special borderline cases can we retain it. But if we pursue the line of thought indicated above, we shall find what it is that we can substitute for the concept of the particle in more general cases.

[The following brief section may be omitted by readers not interested in the somewhat technical issues, and the subject resumed on p. 20.]

[The Quantum Theory postulates that an equation subsists between energy and frequency. If this postulate is to have an unambiguous meaning, that is a meaning independent of the particular system to which it is referred, then the Principle of Relativity demands that a momentum vector[3] shall be equivalent to a wave-member vector; in other words, the

---

[3]A vector is a quantity which has a definite direction; for example, "100 miles per hour East" (or any other direction) is a vector. [TRANS.]

absolute quantity of the momentum must be equivalent to the reciprocal of the length of a wave whose normal coincides with the direction of momentum. The wave in question must not be imagined as existing in ordinary three-dimensional space, but in so-called configuration space, the dimension of which is given by the number of degrees of freedom of the system, and in which the square of the element of length is measured by twice the kinetic energy; or what comes to the same thing, by the square of the total momentum. It thus appears that the wave-length follows from the kinetic energy, that is from the difference between the constant total energy and the potential energy; this difference must be regarded as a function of position given beforehand.

The product of the frequency and the wavelength gives us the rate of propagation of the wave; in other words, it gives us the phase velocity of a given wave the so-called material wave in configuration space. If the appropriate values are substituted in the familiar equation of classical mechanics, we obtain the linear homogeneous partial differential equation set up by Schrödinger. This equation has provided the basis of modern Quantum mechanics, in which it seems to play the same part as do the equations established by Newton, Lagrange and Hamilton in classical mechanics. Nevertheless there is an important distinction between these equations, consisting in the fact that in the latter equations the coordinates of the configuration point are not functions of time, but independent variables. Accordingly, while for any given system the classical equations of motion were more or less numerous and corresponded to the number of degrees of freedom of the system, there can be only one single quantum-equation for each system. In course of time the configuration point of classical theory describes a definite curve; on the other hand, the configuration point of the material wave fills at any given time the whole of infinite space, including those parts of space where potential energy is greater than the total energy, so that according to the classical theory, kinetic energy would become negative in these parts of space, and the momentum imaginary.

This case resembles the so-called total reflection of light, where according to geometrical optics light is completely reflected, because the angle of refraction becomes imaginary; whereas according to the wave-theory of light, it is perfectly possible for light to penetrate into the second medium, even if it cannot do so as a plane wave.

At the same time, the fact that there are points in configuration space where the potential energy exceeds the total energy is of extreme importance for Quantum mechanics. Calculation shows that in every such instance a finite wave corresponds not to any given value of the energy constant, but corresponds only to certain definite values: the so-called characteristic energy-values, which can be calculated from the wave-equation and have different values according to the nature of the given potential energy.

From the discrete characteristic energy-values, discrete characteristic values of the period of oscillation may be derived. The latter are determined according to the Quantum postulate, in a similar manner to that of a stretched cord with fixed ends; with this distinction that the latter quantization is determined by an external condition, viz. the length of the cord, whereas in the present instance it depends upon the Quantum of Action, which in turn depends directly upon the differential equation.

To each characteristic vibration there corresponds a particular wave-function (ψ); this is the solution of the wave-equation; and all these different characteristic functions form the component elements for the description of any movement in terms of wave-mechanics.

Thus we reach the following result: in classical Physics the physical system under consideration is divided spatially into a number of smallest parts; by this means the motion of material bodies is traced back to the motion of their component particles, the latter being assumed to be unchangeable. In other words, the explanation is based upon a theory of corpuscles. Quantum Physics, on the other hand, analyses all motion into individual and periodic material waves, which are taken to correspond to the characteristic vibrations and characteristic functions of the system in question; in this way it is based upon wave-mechanics. Accordingly, in classical mechanics the simplest motion is that of an individual particle, whereas in quantum-mechanics the simplest motion is that of a simple periodic wave; according to the first, the entire motion of a body is taken as being the totality of the motions of its component particles; whereas according to the second, it consists in the joint effect of all kinds of periodic material waves. To illustrate the difference between these two views, we may once more refer to the vibrations of a stretched cord. On the one hand these vibrations may be imagined as consisting of the sum of the motions of the different particles of the cord, where each particle is in motion independently of all the rest and in accordance with the force acting upon it, which in turn depends upon the local curvature of the cord. On the other hand the process of vibration may be analysed into the fundamental and upper partial vibrations of the cord, where each vibration affects the cord in its totality and the sum total of vibration is the most general kind of motion taking place in the cord.

Wave-mechanics also furnishes an explanation for another fact which hitherto has been inexplicable. According to Niels Bohr's theory, the electrons of an atom move around the nucleus in accordance with laws very similar to those which govern the motion of the planets around the sun. Here the place of gravitation is taken by the attraction between the opposite charges of the nucleus and the electrons. There is, however, a curious distinction, consisting in the fact that the electrons can move only in definite orbits distinct from each other, whereas with the planets no one orbit appears to be privileged beyond any other. According to the wave theory of electrons this circumstance, at first sight unintelligible, is easily explained. If the orbit of an electron returns upon itself, it is clear that it must comprise an integral number of wavelengths, just as the length of a chain which forms a complete circle, if it consists of a number of equal links, must always equal an integral number of link-lengths. According to this view the revolution of an electron around the nucleus is not so much like the movement of a planet around the sun, as like the rotation of a symmetrical ring upon its centre, so that the ring as a whole retains the same position in space; thus there is no physical meaning in referring to the local position of the electron at any instant.

The following question may now be asked: if motion is to be analysed not into particles, but into material waves, what is the procedure of wave-mechanics when it is called upon to describe the motion of a single particle which occupies a given position at a given time? The answer to this question throws light upon the great contrast between the two theories with which we have been dealing. In the first instance we must examine the physical meaning of the wave function ψ of a simple periodic material wave. This meaning can be derived from the consideration that the energy of a material wave has a twofold meaning. It is true that it denotes the period of vibration of the wave; but of course it does not follow from this that it has lost its original meaning, which it derives from the principle of conservation of energy. But if the energy principle is to apply to wave-mechanics, then it must be possible to represent the energy of a material wave, not only by the frequency of its vibrations, but also by means of an integral comprehending the entire configuration space of the wave.

In fact, then, if the wave-equation is multiplied by $\bar{\psi}$ and the product is integrated over the entire configuration space, there results a definite expression for the energy, which can be most vividly interpreted in the following manner.

We imagine the material system of particles under consideration to be multiplied many times, and we further imagine that each of the resulting systems is in a different configuration, so that we obtain a very great number of particles in configuration space. We further allot to the configuration points existing in the different infinitely small elements of space a definite energy which is composed (a) of the value of the local potential energy (which is given beforehand) and (b) of a second element which varies as the square of the local gradient of ψ, and which we can interpret as being equivalent to kinetic energy. If, then, the spatial density of the configuration points at any one place is assumed to be equal to the square of the absolute value of ψ (which latter we may assume to have any magnitude we desire, since one of the constant factors of ψ can be selected by ourselves at will), it follows that the mean energy of all the configuration points is equivalent to the energy of the material wave. Accordingly the absolute value of the amplitude of the wave has no meaning whatever in a physical sense. If we imagine ψ to be selected in such a way that the square of the absolute value of ψ, when integrated over the configuration space, gives us the value 1, then we can also say that this square denotes the probability that the material system of particles is actually existing at the point in question within the configuration space. Thus we have found a vivid expression for the physical meaning of the wave-function ψ, which we were looking for.

In the course of all these considerations we had assumed that ψ had a definite characteristic function of its own, and that there was a simple periodic wave corresponding to it. Similar statements, however, may be made for the general case where waves having different periods are superimposed. In that case the wave-function ψ is the algebraic sum of the periodic characteristic functions multiplied by a certain amplitude constant, and once again the square of the absolute value of ψ denotes the probability for the corresponding position of the configuration point. In the general case, of course, we can no longer speak of one single definite period of vibration of the material waves; on the other hand, however, we can still speak of a definite energy. Accordingly the Quantum-equation $E = h\nu$ loses its original meaning and only gives us an average frequency ν. It is worth noting here that if a sufficiently large number of different waves having approximately equal frequencies are superimposed, the wave-function of the resulting wave is the sum of the individual wave-functions; its energy on the other hand does not increase proportionately with the number of individual waves, but always retains its original mean value; the energy of a group of individual waves defines a mean frequency, and similarly the momentum of this group serves to define a mean wave-length.

To begin with, the amplitudes and phases of the individual waves can be selected at will. Beyond this, however, it is impossible to introduce further variety into the mechanical processes of which wave-mechanics can provide instances. This fact becomes important when we turn to the question raised above, in which we ask how the motion of a single definite particle is to be described in terms of wave-mechanics. It appears immediately that *such a description cannot be made in any exact sense.* Wave-mechanics possesses only one means of defining the position of a particle, or more generally the position of a definite point in configuration space; this consists in superimposing a group of individual waves of the system, in such a manner that their wave-functions cancel each other by interference everywhere within configuration space, and intensify each other only at the one point in question. In this case the probability of all the other configuration points would be equal to 0, and would be equal to 1 only for the one point in question. In order to isolate this point completely we should, however, require infinitely small wave-lengths, and consequently infinitely great momentum. Therefore, in order to obtain a result which would be even approximately useful, we should have to begin by substituting for the definite configuration point a finite (though still small) region of configuration space, or so-called wave-group; this sufficiently expresses the fact that ascertaining the position of a configuration point is always in the wave theory affected by some sort of uncertainty.

If we wish to go further and ascribe to the system of particles a definite quantity of momentum as well as a definite configuration, then the Quantum postulate, if taken strictly, will allow us to make use of only one single wave of a definite length for our exposition, and once more description becomes impossible. On the other hand, if a slight uncertainty is allowed to creep into the quantity of momentum, then we can reach our goal, at least approximately, if we make use of the wave within a certain narrow range of frequency.

According to wave-mechanics, both the position and the momentum of a system of particles can never be defined without some uncertainty. Now the fact is that between these two kinds of uncertainty there is a definite relation. This follows from the simple reflection that if the waves of which we make use are to cancel each other through interference outside the above-mentioned small configuration region, then in spite of their small difference in frequency, noticeable differences in propagation must appear at the opposite boundaries of the region. If in accordance with the Quantum postulate, we substitute differences of momentum for differences of propagation, we obtain Heisenberg's Principle, which states that the product of the uncertainty of position and uncertainty of momentum is at least of the same order of magnitude as the quantum of action.]

The more accurately the position of the configuration point is ascertained, the less accurate is the amount of momentum; and conversely. These two kinds of uncertainty are thus in a certain sense complementary; this complementariness is limited by the fact that momentum can under certain conditions be defined with absolute accuracy in wave-mechanics, whereas the position of a configuration point always remains uncertain within a finite region.

Now this relation of uncertainty, established by Heisenberg, is something quite unheard of in classical mechanics. It had always been known, of course, that every measurement is subject to a certain amount of inaccuracy; but it had always been assumed that an improvement in method would lead to an improvement in accuracy, and that this process could be carried on indefinitely. According to Heisenberg, however, there is a definite limit to the accuracy obtainable. What is most curious is that this limit does not affect position and velocity separately, but only the two when combined together. In principle, either taken by itself can be measured with absolute accuracy, but only at the cost of the accuracy of the other.

Strange as this assertion may seem, it is definitely established by a variety of facts. I will give one example to illustrate this. The most direct and accurate means of ascertaining the position of a particle consists in the optical method, when the particle is looked at with the naked eye or through a microscope, or else is photographed. Now for this purpose the particle in question must be illuminated. If this is done the definition becomes more accurate; consequently the measurement becomes more exact in proportion as the light-waves employed become shorter and shorter. In this sense, then, any desired degree of accuracy can be attained. On the other hand there is also a disadvantage, which affects the measurement of velocity. Where the masses in question have a certain magnitude, the effect of light upon the illuminated object may be disregarded. But the case is altered if a very small mass, e.g. a single electron, is selected; because each ray of light, which strikes the electron and is reflected by it, gives it a distinct impulse; and the shorter the light-wave the more powerful is this impulse. Consequently, the shorter the light-wave the more accurately is it possible to determine position; but at the same moment measurement of velocity becomes proportionately inaccurate; and similarly in analogous instances.[4]

On the view which has just been set out classical mechanics, which is based on the assumption of unchanging and accurately measurable corpuscles moving with a definite velocity, forms one ideal limiting case. This ideal case is actually realized when the observed system possesses a relatively considerable energy. When this happens, the distinct characteristic energy values will lie close to each other, and a relatively small region of energy will contain a considerable number of high wave-frequencies (i.e. of short wave-lengths); through the superposition of these a small wave-group with definite momentum can be delimited comparatively accurately within the configuration space. In this case, wave-mechanics merges with the mechanics of particles; Schrödinger's differential equation becomes the classical differential equation of Hamilton and Jacobi, and the wave-group travels in configuration space in accordance with the same laws which govern the motion of a system of particles according to classical mechanics. But this state of affairs is of a limited duration; for the individual material waves are not interfering continually in the same manner, and consequently the wave-group will disintegrate more or less quickly; the position of the relative configuration point will become more and more uncertain, and finally

---

[4]EDITOR'S NOTE: This illustration of Heisenberg's relation of uncertainty is now regarded as inaccurate and misleading, because it implies that it is the very process of measurement that gives rise to this relation, whereas the relation of uncertainty is an intrinsic feature of quantum phenomena regardless of whether they are measured or not. In fact, even Planck himself points out that "The principle [Heisenberg's Principle of Uncertainty] has nothing whatever to do with any measurement" (p. 28).

the only quantity remaining that is accurately defined is the wave-function ψ.

The question now arises whether these conclusions correspond with experience. Since the Quantum of Action is so small, this question can be answered only within the framework of atomic physics; consequently the methods employed will always be extremely delicate. At present we can only say that hitherto no fact has been discovered which throws doubt on the applicability in Physics of all these conclusions.

The fact is that since the wave-equation was first formulated, the theory has been developing at a most remarkable rate. It is impossible within the framework of a small volume to mention all the extensions and applications of the theory which have been evolved within recent years. I shall confine myself to the so-called stress of protons and electrons; the formulation of Quantum mechanics in terms of Relativity; the application of the theory to molecular problems, and the treatment of the so-called "many-body problem", i.e. its application to a system containing a number of exactly similar particles. Here statistical questions, relating to the number of possible states within a system, having a given energy, are particularly important; they also have a bearing on the calculation of the entropy of the system.

Finally, I cannot here enter in detail upon the Physics of light-quanta. In a certain sense this study has developed in the opposite direction from the Physics of particles. Originally Maxwell's theory of electromagnetic waves dominated this region, and it was not seen until later that we must assume the existence of discrete light particles; in other words that the electromagnetic waves, like the material waves, must be interpreted as waves of probability.

Perhaps there is no more impressive proof of the fact that a pure wave theory cannot satisfy the demands of modern Physics any more than a pure corpuscular theory. Both theories, in fact, represent extreme limiting cases. The corpuscular theory, which is the basis of classical mechanics, does justice to the configuration of a system, but fails to determine the values of its energy and of momentum; conversely the wave theory, which is characteristic of classical electrodynamics, can give an account of energy and momentum, but excludes the idea of the localization of light particles. The standard case is represented by the intermediate region, where both theories play equally important parts; this region can be approached from either side, although at present a close approach is impossible. Here many obscure points await solution, and it remains to be seen which of the various methods employed for their solution best leads to the goal. Among them we may mention the matrix calculus invented by Heisenberg, Born, and Jordan, the wave theory due to de Broglie and Schrödinger, and the mathematics of the q numbers introduced by Dirac.

§ 4

If we attempt to draw a comprehensive conclusion from the above description and to obtain an insight into the distinguishing characteristics of our new picture of the world, the first impression will no doubt be somewhat unsatisfactory. First of all it will appear surprising that wave-mechanics, which itself is in complete contradiction to classical mechanics, nevertheless makes use of concepts drawn from the classical corpuscular theory; e.g. the concept of the coordinates and momentum of a particle, and of the kinetic and potential energy of a system of particles. The contradiction is the more surprising since it afterwards proved impossible simultaneously to determine exactly the position and momentum of a particle. At the same time these concepts are absolutely essential to wave-mechanics; for without them it would be impossible to define configuration space and ascertain its measurements.

There is another difficulty attached to the wave theory, consisting in the fact that material waves are not as easy to bring before the imagination as are acoustic or electromagnetic waves; for they exist in configuration space instead of ordinary space, and their period of vibration depends on the choice of the physical system to which they belong. The more extensive this system is assumed to be, the greater will be its energy, and with this the frequency.

It must be admitted that these are serious difficulties. It will be possible, however, to overcome them if two conditions are fulfilled: the new theory must be free from internal contradictions; and its applied results must be definite and of some significance for measurement. At the present time opinions are somewhat divided whether these requirements are fulfilled by Quantum mechanics, and if so, to what extent. For this reason I propose to discuss this fundamental point further.

It has frequently been pointed out that Quantum mechanics confines itself on principle to magnitudes and quantities which can be observed, and to questions which have a meaning within the sphere of Physics. This observation is correct; but in itself it must not be considered a special advantage of the Quantum Theory as opposed to other theories. For the question whether a physical magnitude can in principle be observed, or whether a certain question has a meaning as applied to Physics, can never be answered *a priori*, but only from the standpoint of a given theory. The distinction between the different theories consists precisely in the fact that according to one theory a certain magnitude can in principle be observed, and a certain question have a meaning as applied to Physics; while according to the other theory this is not the case. For example, according to the theories of Fresnel and Lorentz, with their assumption of a stationary ether, the absolute velocity of the earth can in principle be observed; but according to the Theory of Relativity it cannot; again, the absolute acceleration of a body can be in principle observed according to Newtonian mechanics, but according to Relativity mechanics it cannot.[5] Similarly the problem of the construction of a *perpetuum mobile* had a meaning before the principle of the conservation of energy was introduced, but ceased to have a meaning after its introduction. The choice between these two opposed theories depends not upon the nature of the theories in themselves, but upon experience. Hence it is not sufficient to describe the superiority of Quantum mechanics, as opposed to classical mechanics, by saying that it confines itself to quantities and magnitudes which can in principle be observed, for in its own way this is true also of classical mechanics. We must indicate the particular magnitudes or quantities which, according to Quantum mechanics, are or are not in principle observed; after this has been done it remains to demonstrate that experience agrees with the assertion.

Now this demonstration has in fact been completed, e.g. with respect to Heisenberg's Principle of Uncertainty, so far

---

[5] EDITOR'S NOTE: This statement is incorrect, which is puzzling given that, undoubtedly, Planck had been aware of Minkowski's explanation of why acceleration is absolute – because it reflects an absolute geometric feature: the *curvature* (deformation) of a worldline. Minkowski explained that a particle which moves with constant velocity is a straight timelike worldline in spacetime, whereas an accelerated particle (accelerating along a straight line or rotating) is a curved (rather deformed) timelike worldline. And he pointed out that "Especially the concept of acceleration acquires a sharply prominent character" (Hermann Minkowski, *Space and Time: Minkowski's papers on relativity* (Minkowski Institute Press, Montreal 2012), p. 117). Now it is clear that acceleration is absolute in both flat and curved spacetime (i.e. in special and general relativity) for the same reason revealed by Minkowski – because a non-accelerating particle is represented by a geodesic worldline in spacetime, whereas the worldline of an accelerating particle is deformed; as Minkowski explained, deformation of a worldline (the deviation of its geodesic shape) is an absolute geometric feature explaining the absolute physical feature of particles - their *resistance* to being accelerated (resistance is an absolute physical feature because it is an experimental fact).

as seems possible at the present moment, and to this extent it can be looked upon as proving the superiority of wave-mechanics.

In spite of these considerable successes, the Principle of Uncertainty which is characteristic of Quantum Physics has caused considerable hesitation, because the definition of magnitudes and quantities which are continually in use is in principle treated as being inexact by this theory. This dissatisfaction is increased by the fact that the concept of probability has been introduced in the interpretation of the equations used in Quantum mechanics; for this seems to imply a surrender of the demands of strict causality in favour of a form of indeterminism. Today, indeed, there are eminent physicists who under the compulsion of facts are inclined to sacrifice the principle of strict causality in the physical view of the world.

If such a step should actually prove necessary the goal of physicists would become more remote; and this would be a disadvantage whose importance it is impossible to overestimate. For in my opinion, so long as any choice remains, determinism is in all circumstances preferable to indeterminism, simply because a definite answer to a question is always preferable to an indefinite one.

So far as I can see, however, there is no ground for such a renunciation. For there always remains the possibility that the reason why it is impossible to give a definite answer resides, not in the nature of the theory, but in the manner in which the question is asked. If a question is inadequately formulated physically, the most perfect physical theory can give no definite answer; a fact widely known in classical statistics and frequently discussed. For example, if two elastic spheres strike one another in a plane, while their velocities before impact and the laws of impact are known in all their details, it still remains impossible to state their velocities after impact. The fact is that, in order to calculate the four unknown components of the velocities of the two spheres after impact,

we have only three equations derived from the conservation
of energy and the two components of momentum. From this,
however, we do not infer that there is no causality governing
impact phenomena; what we do say is that certain essential
data are missing which are requisite for their complete determination.

In order to apply these considerations to the problems of
Quantum Physics, we must now return to the arguments dealt
with in the Introduction.

If it is really true that, in its perpetual changes, the structure of the physical world-view moves further and further
away from the world of the senses, and correspondingly ap-
proaches the real world (which, as we saw, cannot in principle
be apprehended at all), then it plainly follows that our view
of the world must be purged progressively of all anthropomorphic elements. Consequently we have no right to admit
into the physical world-view any concepts based in any way
upon human mensuration. In fact this is not the case with
Heisenberg's Principle of Uncertainty: this was reached from
the consideration that the elements of the new view of the
world are not material corpuscles, but simple periodic material waves which correspond to the physical system under
consideration – a conclusion obtained in accordance with the
mathematical principle that it is impossible to determine a
definite particle with definite momentum by means of superposition of simple periodic waves having a finite length. The
principle has nothing whatever to do with any measurement,
while the material waves are definitely determined by means
of the mathematical problem of boundary values relating to
the case in question. Here there is no question of indeterminism.

The question of the relation between the material waves
and the world of sense is a different one. For this relation
renders it possible for us to become acquainted with physical
events; if a system were completely cut off from its surround-

ings we could never know of its existence.

At first glance it appears that this question has nothing to do with Physics, since it belongs partly to Physiology and partly to Psychology. These objections, however, lead to no real difficulty. It is always possible to imagine suitably constructed instruments being substituted for human senses, e.g. self-registering apparatus like a sensitive film, which registers the impressions derived from the environment, and is thus capable of furnishing evidence about the events taking place in these surroundings. If such instruments are included within the physical system which we propose to consider, and if all other influences are eliminated, then we have a physical system cut off from the rest of the world of which we can discover something by means of measurement; although it is true that we must take into account the structure of the measuring instruments, and the reaction which they might conceivably have upon the events which we desire to measure.

If we possessed an instrument reacting to a simple periodic material wave in the same way as a resonator reacts to a sound-wave, then we would be in a position to measure individual material waves and thus to analyse their behaviour. This is not the case; the fact is that the indications given by such instruments as we possess, e.g. the darkening of a photographic film, do not allow us to make a safe inference about all the details of the process under examination. We have no right, however, to infer from this that the laws of material waves are indeterminate.

Another and more direct attempt might be made to substantiate the assumption of indeterminism from the fact that, according to wave-mechanics, the events within a system of particles cut off from the outside world are not determined in any way by the initial state of the system, i.e. by the initial configuration and initial momentum. There is not even an approximate determination; for the wave-group corresponding to the initial state will in time disintegrate generally and fall apart into individual waves of probability.

On closer consideration, however, we see that in this instance the element of indeterminism is due to the manner in which the question is asked. The question is based upon corpuscular mechanics; and in corpuscular mechanics the initial state governs the course of the event for all time. But in wave-mechanics such a question has no place, if only because the final result is on principle affected with a finite inaccuracy due to the Principle of Uncertainty.

Since the times of Leibniz, on the other hand, another form of question in classical mechanics has been known which in this sphere leads to a definite answer. An event is completely determined for all time if, apart from the configuration at a certain time, we know, not the momentum, but the configuration of the same system at a different instant. In this case the principle of variation, or principle of least action, is used in order to calculate the event. To take the previous example, where two elastic spheres meet in a plane, if we know the initial and final position of the spheres and the interval between those two positions, then the three unknown quantities, namely the two local coordinates and the time coordinate of impact, are completely determined by the three equations of conservation.

This changed formulation of the problem differs from the previous formulation because it is immediately applicable to wave-mechanics. It is true, as we saw, that a given configuration can never be defined with complete accuracy by the wave theory; but on the other hand it is theoretically possible to reduce the uncertainty below any desired limit, and thus to determine the event in question with any desired degree of accuracy. Further, the disintegration of wave-groups is no evidence in favour of indeterminism, since it is equally possible for a wave-group to conglomerate: in both the wave theory and the corpuscular theory the *direction* of the process is immaterial. Any movement might equally well take place in the opposite direction.

When the above formulation of the problem is adopted a given wave-group generally, of course, exists only at the two selected instants: in the intervening period, as well as before and after the process, the different elementary waves will exist separately. But whether they are described as material waves or as waves of probability, in either case they will be completely determined. This is the explanation of the apparent paradox, that when a physical system passes by a definite process from one definite configuration during a definite time into some other definite configuration, the question what its configurations are during the intervening period has no physical significance; similarly on this view there is no meaning in the question of what is the track of light quantum emitted from a point source and absorbed at a given point on an observation-screen.

It should at the same time be emphasized that on this view the meaning of determinism is not exactly what it is in classical Physics. In the latter the configuration is determined; in Quantum Physics, the material waves. The distinction is important, because the connection between the configuration and the world of sense is far more direct than that between the material waves and the sense-world. To this extent the relation between the physical world-view and the world of sense appears to be considerably looser in modern Physics.

This is undoubtedly a disadvantage; but it is the price that must be paid in order to preserve the determinism of our world-view. And further, this step appears to lie in the general direction in which Physics is actually developing; this has been pointed out on more than one occasion, since in the course of its progressive evolution, the structure of the physical view of the world is moving farther and farther away from the world of sense, and assuming more and more abstract forms. Indeed, the principle of Relativity seems actually to demand such a view; for on this principle Time stands on the same level with Space, whence it follows that, if a finite space is required for the causal description of a physical process, a finite temporal interval must also be used in order to complete the description.

On the other hand, it may well be that the suggested formulation of the question is too one-sided, and too anthropomorphic to furnish satisfactory material for a new theory of the structure of the physical world; it may be that we shall have to look for some other formulation. In any case many complex problems remain to be solved, and many obscure points to be cleared up.

In view of the peculiar difficulties of the position which has been reached by theoretical Physics, a feeling of doubt persists whether the theory, with all its radical innovations, is really on the right path. The answer to this decisive question depends wholly upon the degree of necessary contact with the sense world which the physical world-view maintains in the course of its incessant advance. If this contact is lost even the most perfect world-view would be no better than a bubble ready to burst at the first puff of wind. There is, fortunately, no cause for apprehension, at least in this respect: indeed we may assert without exaggeration that there was no period in the history of Physics when theory and experience were linked so closely together as they are now. Conversely, it was the facts learned from experiments that shook and finally overthrew the classical theory. Each new idea and each new step were suggested to investigators, where it was not actually thrust upon them, as the result of measurements. The Theory of Relativity was led up to by Michelson's experiments on optical interference, and the Quantum Theory by Lummer's, Pringsheim's, Ruben's and Kurlbaum's measurements of the spectral distribution of energy, by Lenard's experiments on the photoelectric effect, and by Franck and Hertz's experiments on the impact of electrons. It would lead me too far if I were to enter on the numerous and surprising results which have compelled Physical theory to abandon the classical standpoint and to enter on a definite new course.

We can only hope that no change will take place in this peaceful international collaboration. It is in this reciprocal action of experiment and theory – which is at once a stimulus to and a check upon progress – that we see the surest and indeed the only guarantee of the future advance of Physics.

What will be the ultimate goal? I had occasion at the beginning to point out that research in general has a twofold aim – the effective domination of the world of sense, and the complete understanding of the real world; and that both these aims are in principle unattainable. But it would be a mistake to be discouraged on this account. Both our theoretical and practical tangible results are too great to warrant discouragement; and every day adds to them. Indeed, there is perhaps some justification for seeing in the very fact that this goal is unattainable, and the struggle unending, a blessing for the human mind in its search after knowledge, For it is in this way that its two noblest impulses – enthusiasm and reverence – are preserved and inspired anew.

§ 5

What now do we mean by physical law? A physical law is any proposition enunciating a fixed and absolutely valid connection between measurable physical quantities – a connection which permits us to calculate one of these quantities if the others have been discovered by measurement. The highest and most keenly desired aim of any physicist is to obtain the most perfect possible knowledge of the laws of Physics, whether he looks at them from a utilitarian point of view and values them because they enable him to save himself the trouble of costly measurements, or takes a deeper view and looks to them for satisfaction of a profound yearning after knowledge and for a firm basis of natural science.

How do we discover the individual laws of Physics, and what is their nature? It should be remarked, to begin with, that we have no right to assume that any physical laws exist, or if they have existed up to now, that they will continue to exist in a similar manner in future. It is perfectly conceivable that one fine day Nature should cause an unexpected event to occur which would baffle us all; and if this were to happen we would be powerless to make any objection, even if the result would be that, in spite of our endeavours, we should fail to introduce order into the resulting confusion. In such an event, the only course open to science would be to declare itself bankrupt. For this reason, science is compelled to begin by the general assumption that a general rule of law dominates throughout Nature, or, in Kantian terminology, to treat the concept of causality as being one of the categories which are given *a priori* and without which no kind of knowledge can be attained.

From this it follows that the nature of the laws of Physics, and the content of these laws, cannot be obtained by pure thought; the only possible method is to turn to the investigation of Nature, to collect the greatest possible mass of varied experiences, to compare these and to generalize them in the simplest and most comprehensive proposition. In other words, we must have recourse to the method of induction.

The content of an experience is proportionally richer as the measurements upon which it is based are more exact. Hence it is obvious that the advance of physical knowledge is closely bound up with the accuracy of physical instruments and with the technique of measurement. The latest developments of Physics provide us with striking examples of the truth of this. Measurement alone, however, does not suffice. For each measurement is an individual event standing by itself; as such, it is determined by special circumstances, especially by a definite place and a definite time, but also by a definite measuring instrument, and by a definite observer. It is true that frequently the generalization which is our object is quite obvious and, so to speak, thrusts itself upon us; on the other hand, there are cases where it is extremely difficult to find the common law governing a number of different measurements, either because it seems impossible to find such a law, or because a number of different laws seem available in order to generalize the facts. Both possibilities are equally unsatisfactory.

In such cases, the only method of advance consists in introducing a so-called working hypothesis to see what it is worth and how far it will lead. It is generally a sign that the hypothesis is likely to turn out useful if it works even in those regions for which it was not originally designed. In such a case we have a right to assume that the law which it enunciates has a deeper meaning and opens the way to unmistakably new knowledge.

We see then that a good working hypothesis is essential for inductive investigation. This being so, we are faced with the difficult question how we are to set about to find the most suitable hypothesis. For this there can be no general rule. Logical thought by itself does not suffice – not even where it has an exceptionally large and manifold body of experience to aid it. The only possible method consists in immediately gripping the problem or in seizing upon some happy idea. Such an intellectual leap can be executed only by a lively and independent imagination and by a strong creative power, guided by an exact knowledge of the given facts so that it follows the right path.

Such an intellectual process generally consists in the introduction of certain mental images and analogies which point the way to the reigning laws already known in other regions, thus suggesting a further step towards the simplification of the physical view of the world.

It is precisely at these points where success seems to be awaiting us, however, that a serious danger is frequently hidden. Once a step forward has succeeded and the working hypothesis has demonstrated its usefulness, we must go further. We have to reach the actual essence of the hypothesis and, by suitably formulating it, we have to throw light upon its genuine content by eliminating everything that is inessential. This process is not as simple as it might appear. The intellectual leap of which we spoke above constructs a kind of bridge by which we can approach fresh knowledge; but on closer examination it frequently appears that this bridge is merely provisional, and that a more enduring structure must be put in its place, capable of bearing the heavy artillery of critical logic.

We must bear in mind that every hypothesis is the outcome of the efforts of imagination, and that imagination works through direct intuition. But in Physics, as soon as we come to look for a rational theory or a logical demonstration, direct intuition is a very doubtful ally, however indispensable it may be while we are forming our hypothesis. For while it is natural that we should rely upon imaginations and ideas of this kind, which proved fruitful in one direction or another, such reliance is only too apt to lead to an overestimation of their importance and to untenable generalizations. We must further recognize that the authors of a new and practicable theory are frequently little inclined to introduce any important changes in the groups of ideas which led them towards their discoveries, whether from indolence or from a certain sentimental feeling, and that they often exert the whole of their well-earned authority in order to be able to maintain their original standpoint. Thus we shall readily understand the difficulties which often stand in the way of healthy theoretical development. Examples may be found at every point in the history of Physics, and I propose to enumerate some of the more important of them.

The first exact measurements were made in the region of Space and Time – the first region where accurate measurement was possible. Hence naturally the earliest physical laws were discovered in this field; in other words, in the sphere of mechanics. Again, we can readily understand how it came about that the first laws which were established related to those motions which occur regularly and independently of external interference, namely, the motions of the celestial bodies. We know that the civilized peoples of the East had discovered thousands of years ago how to derive from their observations formulae which allowed them to calculate in advance the motion of the sun and the planets with great accuracy. Each improvement in the instruments of measurement was accompanied by an improvement of the formulae. By their coordination and comparison the theories of Ptolemy, Copernicus and Kepler were evolved in course of time, each of which is simpler and more exact than those which preceded it. All these theories are alike in endeavouring to answer the question, what is the connection between the position of a celestial body, a planet for example, and the moment of time at which it occupies this position? The nature of this necessary connection is, of course, different for the different planets, and this in spite of the fact that the motions of the planets have many characteristics in common.

The decisive step beyond this type of question was taken by Newton. Newton summed up all the formulae relating to the planets in one single law governing their motion, and indeed that of all the celestial bodies. He was enabled to do this because he made the law of motion independent of the particular moment to which it is applied: for the instant he substituted the time-differential. Newton's theory of planetary motion enunciates a fixed connection not between the position of a planet and time, but between the acceleration of a planet and its distance from the sun. Now this law – a vectorial differential equation – is the same for all the planets. Hence if the position and velocity of a planet are known for any moment, then its motion for all time can be exactly calculated.

The successes obtained as the result of the further application of Newton's formulation of the laws of motion prove that it is not merely a new description of certain natural phenomena, but that it represents a real advance in the understanding of actual facts. It is not merely more exact than Kepler's formulae, for example when it allows for the interference in the elliptical orbit of the earth around the sun due to the periodic proximity of Jupiter, where formula and measurement are in exact agreement; more than that, it also covers the motion of such bodies as comets, twin stars, etc., which altogether elude Kepler's laws. The complete and immediate success of Newton's theory was due, however, to the fact that when applied to motion occurring on the earth, it led to the same numerical laws of gravitation and pendulum movements which Galileo had already discovered by measurement, and also threw light on otherwise inexplicable phenomena, such as those of tides, rotation of the plane of the pendulum, precession of the axis of rotation, etc.

The question which especially interests us at the moment is how Newton reached his differential equation for planetary motion. He did not reach it by establishing a connection between the acceleration of a planet and its distance from the sun, and by looking for a numerical connection between them; what he did was first to forge an intellectual link between them, leading from the concept of the position of a planet to that of its acceleration; and this link he called Force. He assumed that the position of a planet relatively to the sun depends upon a force of attraction directed towards the sun, and that the same attractive force also causes a definite change in the planet's motion. This was the germ of the law of gravitation, as well as of the law of inertia. The notion of force was no doubt derived (as the word implies) from the idea of the muscular sensation which arises when a weight is lifted or a ball is thrown; this idea was generalized and applied to every kind of change of motion, even where the forces in question are so great that no human power could possibly suffice to effect them.

Small wonder, then, that Newton attributed the greatest importance to the concept of force which had helped him to reach such striking results. At the same time it must be noted that this concept does not occur in the law of motion proper. Newton looked to the concept of force for an explanation of every change of motion; and thus it came about that Newtonian force was regarded as the main and fundamental concept in mechanics, and not only in mechanics, but also in Physics; so that, in course of time, physicists formed the habit of making their first question when dealing with physical phenomena: what force is here in action?

Recent developments in Physics present a certain contrast with Newton's theory, so that in a manner it is true to say that the concept of force is no longer of fundamental importance for physical theory. In modern mechanics force is no more than a magnitude of secondary importance, and its place has been taken by higher and more comprehensive concepts – that of work[6] or potential, where force in general is defined as a negative potential gradient.

It might here be objected that work surely cannot be looked upon as something primary, since there must be some kind of force in existence that does the work. This kind of argument is of the physiological and not of the physical order. It is true that in lifting a weight the contraction of the muscles and the accompanying sensations are primary, and are the cause of the motion which actually takes place. But this kind of work, which is a physiological process, must be clearly distinguished from the physical force of attraction with which alone we are here concerned; it is this force which the earth exerts upon everything having weight; and this in its turn

---

[6] "Work" is here used in its scientific sense of the product of the force and the distance through which the force acts. [TRANS.]

depends upon the gravitational potential which is already in existence and is primary.

The idea of potential is superior to that of force, partly because it simplifies the laws of Physics, and also because the significance of the idea of potential has a far greater scope than that of force; it reaches beyond the sphere of mechanics into that of chemical affinities, where we are no longer concerned with Newtonian force. It must be admitted that the idea of potential has not the advantage of immediate obviousness which belongs to force by virtue of its anthropomorphic quality; whence it follows that the elimination of the concept of force renders the laws of Physics much less obvious and easy of understanding. Yet this development is quite natural; the laws of Physics have no consideration for the human senses; they depend upon facts, and not upon the obviousness of facts.

In my opinion, the teaching of mechanics will still have to begin with Newtonian force, just as optics begins with the sensation of colour, and thermodynamics with the sensation of warmth, despite the fact that a more precise basis is substituted later on. Again, it must not be forgotten that the significance of all physical concepts and propositions ultimately does depend on their relation to the human senses. This is indeed characteristic of the peculiar methods employed in physical research. If we wish to form concepts and hypotheses applicable to Physics, we must begin by having recourse to our powers of imagination; and these depend upon our specific sensations, which are the only source of all our ideas. But to obtain physical laws we must abstract exhaustively from the images introduced, and remove from the definitions set up all irrelevant elements and all imagery which do not stand in a logical connection with the measurements obtained. Once we have formulated physical laws, and reached definite conclusions by mathematical processes, the results which we have obtained must be translated back into the language of the world of our senses if they are to be of any use to us. In a manner this method is circular; but it is essential, for the simplicity and universality of the laws of Physics are revealed only after all anthropomorphic additions have been eliminated.

The concept of Force as used by Newton is only one of a number of intellectual links and auxiliary notions employed in order to render an idea more intelligible. In this connection I should like to mention the idea of osmotic pressure introduced by van't Hoff. This idea proved particularly fruitful in physical chemistry, where it was used in order to formulate the physical laws of solutions, especially of the freezing-point and steam-pressure. To obtain instances of osmotic pressure, and measure it accurately, is not altogether easy, since an extremely complex apparatus (the so-called semi-permeable membranes) is required. We must the more admire the intuitive insight which led van't Hoff to formulate the laws known under his name despite the scantiness of the observed facts at his disposal. Yet in their present form these laws require osmotic pressure no more than the laws of motion require Newtonian Force.

Besides the above there are other kinds of intellectual aids which assist imagination, and have proved of great assistance in the formation of working hypotheses, but which in the further course of development actually embarrassed later progress. One of these is particularly worth mention here. Men had accustomed themselves to see in some kind of force the cause underlying every natural change; and thus they were all the more disposed to imagine every invariable and constant magnitude or quantity as being of the nature of a *Substance*. From the earliest times the concept of Substance has played an important part in Physics: but closer examination shows that this has not always been helpful. It is, of course, easy to see that wherever conservation is concerned, it is possible to assume a Substance of which conservation is predicated; and such an assumption undoubtedly makes it easier to grasp the meaning of the principle, and hence facilitates its use. A magnitude which in spite of every change retains its quantity surely cannot be imagined more vividly than in the shape of a moving material body. It is a feature of this tendency that we are so prone to interpret all natural events as being movements of masses of substance – a mechanistic interpretation. For example, the origin and distribution of light were explained by wave-motion in a substantial light ether; the chief laws of optics were described in this manner and found to agree with experience – until the moment came when the mechanistic theory of Substance failed and became lost in unfruitful speculation.

Again, for a time, the concept of Substance proved exceedingly useful as applied to Heat. The careful development of calorimetry, during the first half of the last century, was due in the main to the assumption that an unchanging heat-substance flowed from the warmer into the colder body. When it was shown that in these circumstances the amount of heat can be increased (e.g. by friction) the Substance theory defended itself by appealing to supplementary hypotheses. But although this method helped for some time, it did not avail indefinitely.

In the theory of electricity the dangerous consequences of an exaggerated application of the idea of Substance became obvious at an early stage. Here again the idea of a subtle and quickmoving electrical substance, giving rise to certain manifestations of force, serves admirably in order to render plastic before the mind such principles as that of the invariability of the quantity of electricity, and such subsidiary ideas as those of the electrical current and of the reciprocal action of charged conductors carrying a current. Here again, however, the analogy fails as soon as we have to allow for the fact that this view implies the assumption of the existence of two opposite substances, one positive and one negative, which completely neutralize each other when they are combined. Such an occurrence is at least as unthinkable as the creation of two opposite substances (in the usual sense) out of nothing.

In this way we see that imaginative ideas and their resultant viewpoints must be used with the greatest caution, even when they have proved their value for some length of time, and despite the fact that they are indispensable for physical investigation and have provided the key to new knowledge on innumerable occasions. There is only one sure guide towards further development, and that is measurement, together with any logical conclusions that can be drawn from the concepts attached to this method. All other conclusions, and especially those characterized by their so-called self-evidence, should always be looked upon with a certain suspicion. The validity of a proof dealing with well-defined concepts is to be judged by reason and not by intuition.

§ 6

Up to this point we have been considering the manner in which the knowledge of physical laws is obtained. We will now proceed to examine the content and the essential nature of the laws of Physics in somewhat greater detail.

A physical law is generally expressed in a mathematical formula, which permits us to calculate the temporal succession of the events taking place in a certain physical system under certain definite and given conditions. From this point of view all the laws of Physics can be divided into two main groups.

The first group consists of those laws which remain valid even when the time order is reversed; in other words, when every process that fulfils their requirements can take place in the reversed order without running counter to them. The laws of mechanics and of electrodynamics are of this nature, except in so far as they relate to chemical phenomena and the phenomena of heat. Every purely mechanical or electro-dynamic process can take place in the reverse direction. The movement of a body falling without friction is accelerated in accordance with the same law which governs the retardation of a body rising without friction; the same laws govern the movement of a pendulum to the left and to the right, and a wave can travel equally well in any direction and in any sense; a planet could equally well revolve around the sun westwards as eastwards. The question whether such movements could actually be reversed, and if so under what conditions, is another matter which need not here be discussed: we are now dealing with the law as such, not with the particular facts to which it applies.

The laws belonging to the second group are characterized by the fact that their time order is of essential importance, so that the events taking place in accordance with these laws have only one temporal direction and cannot be reversed. Among these processes we may mention all those in which heat and chemical affinity play a part. Friction is always accompanied by a decrease and never by an increase of relative velocity; where heat is conducted the warmer body always becomes cooler and the cooler body warmer; in diffusion the process invariably leads to a more thorough mixture and not to a progressive separation of the substances in question. Further, these irreversible events always lead to a definite final state; friction to a relative state of rest, the transfer of heat to temperature equilibrium, and diffusion to a completely homogeneous mixture. On the other hand the former class of reversible events knows neither beginning nor end, so long as no interference takes place from outside, but persists in incessant oscillation.

Now if we wish to introduce unity into the physical view of the universe we must somehow find a formula to cover both these contrasted types of law. How is this indispensable result to be brought about? Some thirty years ago theoretical physics was profoundly influenced by the so-called theory of Energetics, which sought to remove the antithesis by assuming that a fall in temperature, for example, was exactly analogous to the fall of a weight or of a pendulum from a higher to a lower position. This theory, however, did not take into consideration the essential fact that a weight can rise as well as fall, and that a pendulum has reached its greatest velocity at the moment when it has attained its lowest position and therefore, by virtue of its inertia, passes the position of equilibrium and moves to the other side. A transference of heat from a warmer to a colder body, on the contrary, diminishes with the diminution of the difference in temperature, while, of course, there is no such thing as any passing beyond the state of temperature equilibrium by reason of some kind of inertia.

In whatever way we look at it, the contrast between reversible and irreversible processes persists; it must therefore be our task to find some entirely new point of view which will allow us to see that after all there is some connection between the different types of laws. Perhaps we shall succeed in showing that one group of laws is a derivative of the other; if so, the question arises which is to be considered the more simple and elementary – the reversible processes or the irreversible.

Some light is thrown on this question by a formal consideration. Every physical formula contains a number of constant magnitudes, together with variable magnitudes which have to be determined by measurement from case to case. The former magnitudes are fixed once for all and give its characteristic form to the functional connection between the variables which is expressed in the formula. Now if we examine these constants more carefully, we shall find that they invariably are the same for the reversible processes, always recurring, however widely different are the attendant outer conditions. Among these are mass, the gravitation constant, the electrical charge and the velocity of light. On the other hand the constants of the irreversible processes, like the capacity for conducting heat, the coefficient of friction and the diffusion constant, depend to a greater or less degree on external circumstance, e.g. temperature, pressure, etc.

These facts naturally lead us to regard the constants of the first group as the simpler, and the laws dependent on them as the more elementary, and to suppose them incapable of further analysis, while treating the constants of the second group, and the laws depending on them, as being of a somewhat more complex nature. In order to test the validity of this assumption we must make our method of investigation somewhat more exact; we must, so to speak, apply a lens of greater power to the phenomena. If the irreversible processes are in fact composite, then the laws governing them can only be roughly valid, so to say; they must be of a statistical nature, since they are valid only for a large scale view or for summary consideration; that is, for the average values resulting from a large number of distinct processes. The more we restrict the number of individual events on which these average values are based, the more plainly will occasional divergences from the general or macroscopic law make themselves felt. In other words, if in fact the view described is correct, then the laws of the irreversible processes, like those of friction, heat distribution and diffusion, must without exception be inexact if looked at microscopically; they must admit of exceptions in individual cases; and these exceptions will be the more striking, the more careful our examination becomes.

Now it so turned out in the course of events that experience tended more and more to confirm this conclusion. This could come about, of course, only as the result of a great improvement in the methods of making measurements. The laws governing the irreversible processes come so very near to being absolutely valid because of the enormous number of individual events of which these processes are composed. If, for example, we take a liquid having the same uniform temperature throughout, then it follows by the general or macroscopic law of the conduction of heat that no heat flows within the liquid. Such however is not precisely the case. For heat is the result of slight and rapid movements of the molecules constituting the liquid; the conduction of heat, consequently, is due to the transference of these velocities when the molecules collide. Hence a uniform temperature does not mean that all the velocities are equal, but that the average value of the velocities for each small quantity of liquid is equal. This quantity in fact comprises a large number of molecules. But if we take a quantity containing a relatively small number of molecules, then the average of their velocities will vary; and the variation will be the greater, the smaller is the quantity of liquid. This principle can nowadays be regarded as a fact fully proved by experiment. One of the most striking illustrations is what is known as the Brownian Movement, which can be observed through the microscope in small particles of powder suspended in liquid. These particles are driven backwards and forwards by the invisible molecules of the liquid; the movement is the more pronounced the higher is the temperature. If we make the further assumption, to which in principle there is no objection, that each individual impulse is a reversible event governed by the strict elementary laws of dynamics, then we may say that the introduction of a microscopic method of examination shows that the laws governing the irreversible processes, or what is the same thing, the laws based upon statistics and mere rough approximation, can be traced back to dynamic, accurate, and absolute laws.

The striking results reached by the introduction of statistical laws in many branches of physical research in recent times have produced a reftiarkable change in the views of physicists. They no longer, as in the earlier days of Energetics, deny or attempt to cast doubt upon the existence of irreversible processes; instead, the attempt is frequently made to place statistical laws in the foreground, and to subordinate to them laws hitherto regarded as dynamic, including even the law of gravitation. In other words, an attempt is made to exclude absolute law from Nature. And indeed, we cannot but be struck by the fact that the natural phenomena which we can investigate and measure can never be expressed by absolutely accurate numbers; for they inevitably contain a certain inaccuracy introduced by the unavoidable defects of measurement itself. Hence it follows that we shall never succeed in determining by measurement whether a natural law is absolutely valid. If we consider the question from the standpoint of the theory of knowledge we come to the same conclusion. For if we cannot even prove that Nature is governed by law (a difficulty which we meet with at the very outset) *a fortiori* we shall be unable to demonstrate that such a law is absolute.

Hence from a logical point of view, we must admit every justification for the hypothesis that the only kind of law in Nature is statistical. It is a different question whether this assumption is expedient in physical research; and I feel strongly inclined to answer this question in the negative. We must consider in the first instance that the only type of law fully satisfying our desire for knowledge is the strictly dynamic type, while every statistical law is fundamentally unsatisfactory, for the simple reason that it has no absolute validity but admits of exceptions in certain cases; so that we are continually faced by the question what these particular exceptional cases are.

Questions of this nature constitute the strongest argument in favour of the extension and further refinement of experimental methods. If it is assumed that statistical laws are the ultimate and most profound type in existence, then there is no reason in theory why, when dealing with any particular statistical law, we should ask what are the causes of the variations in the phenomena? Actually, however, the most important advances in the study of atomic processes are due to the attempt to look for a strictly causal and dynamic law behind every statistical law.

On the other hand, we may discover a law which has always proved absolutely valid within the marginal error due to measurement. In such a case we must admit that it will never be possible to prove by means of measurement that it is not after all of the statistical type. At the same time, it is of great importance whether theoretical considerations induce us to regard the law as being of the statistical, or of the dynamic, type. For in the first case, we should attempt to attain the limits of its validity by means of the continuous refinement of our methods of measurement; in the second case, we should regard such attempts as useless and thus save ourselves much unnecessary labour. So much trouble has already been spent in Physics upon the solution of imaginary problems that such considerations are very far from being irrelevant.

In my opinion, therefore, it is essential for the healthy development of Physics that among the postulates of this science we reckon, not merely the existence of law in general, but also the strictly causal character of this law. This has in fact almost universally been the case. Further, I consider it necessary to hold that the goal of investigation has not been reached until each instance of a statistical law has been analysed into one or more dynamic laws. I do not deny that the study of statistical laws is of great practical importance: Physics, no less than meteorology, geography and social science, is frequently compelled to make use of statistical laws. At the same time, however, no one will doubt that the alleged accidental variations of the climatological curves, of population statistics and mortality tables, are in each instance subject to strict causality; similarly, physicists will always admit that such questions are strictly relevant as that which asks why one of two neighbouring atoms of Uranium exploded many millions of years before the other.

All studies dealing with the behaviour of the human mind are equally compelled to assume the existence of strict causality. The opponents of this view have frequently brought forward against it the existence of free will. In fact, however, there is no contradiction here; human free will is perfectly compatible with the universal rule of strict causality – a view which I have had occasion to demonstrate in detail elsewhere. But as my arguments on this subject have been seriously misunderstood in certain quarters, and since this subject is surely of considerable importance, I propose to discuss it briefly here.

The existence of strict causality implies that the actions, the mental processes, and especially the will of every individual are completely determined at any given moment by the state of his mind, taken as a whole, in the previous moment, and by any influences acting upon him coming from the external world. We have no reason whatever for doubting the truth of this assertion. But the question of free will is not concerned with the question whether there is such a definite connection, but whether the person in question is aware of this connection. This, and this alone, determines whether a person can or cannot feel free. If a man were able to forecast his own future solely on the ground of causality, then and then only we would have to deny this consciousness of freedom of the will. Such a contingency is, however, impossible, since it contains a logical contradiction. Complete knowledge implies that the object apprehended is not altered by any events taking place in the knowing subject; and if subject and object are identical this assumption does not apply. To put it more concretely, the knowledge of any motive or of any activity of will is an inner experience, from which a fresh motive may spring; consequently such an awareness increases the number of possible motives. But as soon as this is recognized, the recognition brings about a fresh act of awareness, which in its turn can generate yet another activity of the will. In this way the chain proceeds, without it ever being possible to reach a motive which is definitely decisive for any future action; in other words, to reach an awareness which is not in its turn the occasion of a fresh act of will. When we look back upon a finished action, which we can contemplate as a whole, the case is completely different. Here knowledge no longer influences will, and hence a strictly causal consideration of motives and will is possible, at least in theory.

If these considerations appear unintelligible – if it is thought that a mind could completely grasp the causes of its present state, provided it were intelligent enough – then such an argument is akin to saying that a giant who is big enough to look down on everybody else should be able to look down on himself as well. The fact is that no person, however clever, can derive the decisive motives of his own conscious actions from the causal law alone; he requires another law – the ethical law, for which the highest intelligence and the most subtle self-analysis are no adequate substitute.

§ 7

Let us however return to Physics, from which these complications are excluded in advance. I propose now to describe the more important characteristics of the cunent view of the physical world. These characteiistics are due to the endeavour to find a strict causal connection, in the manner described above, for all physical processes. A cursory glance suffices to show what changes there have been since the beginning of the century; and we may say that since the days of Galileo and Newton, no such rapid development has ever been known. Incidentally we may point with pride to the fact that German scientists have played an important part in this advance. The occasion of this development was that extreme refinement in measurement which is an essential condition of the progress of science and engineering; in its turn this led to the discovery of new facts, and hence to the revision and improvement of theory. Two new ideas in particular have given modem Physics its characteristic shape. These are laid down in the Theory of Relativity and the Quantum hypothesis respectively; each in its own way is at once fruitful and revolutionary; but they have nothing in common and, in a sense, they are even antagonistic.

For a time Relativity was a universal topic of conversation. The arguments for and against could be heard everywhere – even in the daily Press, where it was championed and opposed by experts and by others who were very far from being experts. Today things have quieted down a little – a state of affairs which is likely to please nobody better than the author of the Theory himself; public interest appears to have become satisfied and to have turned to other popular topics. From this it might perhaps be inferred that the Theory of Relativity no longer plays any part in science. But as far as I can judge, the opposite is the case: for the Theory of Relativity has now become part and parcel of the physical view of the world, and is taken for granted without any further ado. Indeed, novel and revolutionary as was the idea of Relativity (in both the Special and the General form) when first presented to physicists, the fact remains that the assertions it makes and the attacks it delivers were directed not against the outstanding, recognized and approved laws of Physics, but only against certain views which had no better sanction than custom, deeply rooted though they were. These standpoints are of the kind which, as I have already tried to show, afford a suitable basis for a preliminary understanding of the facts of Physics; but they must be discarded as soon as it is found necessary to reach a more general and profound view of the facts.

In this connection the idea of simultaneity is particularly instructive. At first glance, it seems to the observer that nothing could be more obviously true than to say that there is a definite meaning in asserting that two events occurring at two distant points (e.g. on the Earth and on Mars) are simultaneous. Surely every man has a right to traverse great distances timelessly in thought, and to place two events side by side before the mind's eye. Now it must be emphasized that the

Theory of Relativity does not alter this right in any way. If we possess sufficiently accurate measuring instruments, we can determine with complete certainty whether the events are simultaneous; and if the time measurements are accurately made in different ways, and with different instruments which can be used to check each other, the same result will always be obtained. To this extent the Theory of Relativity has brought about no change whatever.

But the Theory of Relativity does not allow us to assume, as a matter of course, that another observer who is moving relatively to ourselves must necessarily regard the two events as simultaneous. For the thoughts and ideas of one person are not necessarily the thoughts and ideas of another. If the two observers proceed to discuss their thoughts and ideas, each will appeal to his own measurements; and when they do this, it will be found that in interpreting their respective measurements they started from entirely different assumptions. Which assumption is correct it is impossible to decide; and it is equally impossible to decide the dispute as to which of the two observers is in a state of rest, and which in a state of motion. This question, however, is of fundamental importance. For the rate of a clock alters while the clock is being moved: a fact which need occasion no surprise; while from this it follows that the clocks of the two observers go at different rates. Thus we reach the conclusion that each can assert with an equal right that he is himself in a state of rest and that his time measurements are correct; and this in spite of the fact that the one observer regards the two events as simultaneous, while the other does not. These ideas and arguments admittedly present a hard task to our powers of imagination; but the sacrifice in clarity is negligible compared with the inestimable advantages which follow from the amazing generality and simplicity of the physical world-view which they render possible.

In spite of this, some readers may still find themselves unable to get rid of the suspicion that the Theory of Relativity contains some kind of internal contradiction. Such readers should reflect that a theory, the entire content of which can be expressed in a single mathematical formula, can no more contain a contradiction than could two distinct conclusions following from the same formula. Our ideas must adjust themselves to the results of the formula and not conversely. Ultimately it is experience that must decide the admissibility and the importance of the Theory of Relativity. Indeed, the fact that experience allows us to test its validity must be looked upon as the most important evidence in favour of the fruitfulness of the theory. Hitherto no instance has been recorded where the Theory conflicts with experience, a fact which I should like to emphasize in view of certain reports which have recently come before the public. Any one who, for whatever reason, considers it possible or probable that a conflict between the Theory and observed facts can be discovered, could do no better than cooperate in extending the Theory of Relativity and in pushing its conclusions as far as possible, since this is the only means of refuting it through experience. Such an undertaking is the less difficult because the assertions made by the Theory of Relativity are simple and comparatively easy to apprehend, so that they fit into the framework of classical Physics without any difficulty.

Indeed, if there were no historical objections I personally would not hesitate for a moment to include the Theory of Relativity within the body of classical Physics. In a manner the Theory of Relativity is the crowning point of Physics, since by merging the ideas of Time and Space it has also succeeded in uniting under a higher point of view such concepts as those of mass, energy, gravitation, and inertia. As the result of this novel view we have the perfectly symmetrical form which the laws of the conservation of energy and of momentum now assume; for these laws follow with equal validity from the Principle of Least Action – that most comprehensive of all physical laws which governs equally mechanics and electrodynamics.

Now over against this strikingly imposing and harmonious structure there stands the Quantum Theory, an extraneous and threatening explosive body which has already succeeded in producing a wide and deep fissure throughout the whole of the structure. Unlike the Theory of Relativity, the Quantum Theory is not complete in itself. It is not a single, harmonious, and perfectly transparent idea, modifying the traditional facts and concepts of Physics by means of a change which, though of the utmost significance in theory, is practically hardly noticeable. On the contrary, it first arose as a means of escape from an *impasse* reached by classical Physics in one particular branch of its studies – the explanation of the laws of radiant heat. It was soon seen, however, that it also solved with ease, or at least considerably helped to elucidate, other problems which were causing unmistakable difficulties to the classical theory, such as photoelectric phenomena, specific heat, ionization, and chemical reactions. Thus it was quickly realized that the Quantum Theory must be regarded, not merely as a working hypothesis, but as a new and fundamental principle of Physics, whose significance becomes evident wherever we are dealing with rapid and subtle phenomena.

Now here we are faced with a difficulty. This does not so much consist in the fact that the Quantum Theory contradicts the traditional views; if that were all, it follows from what has been said that the difficulty need not be taken very seriously. It arises from the fact that in the course of time it has become increasingly obvious that the Quantum Theory unequivocally denies certain fundamental views which are essential to the whole structure of the classical theory. Hence the introduction of the Quantum Theory is not a modification of the classical theory, as is the case with the Theory of Relativity: it is a complete break with the classical theory.

Now if the Quantum Theory were superior or equal to the classical theory at all points, it would be not only feasible but necessary to abandon the latter in favour of the former. This, however, is definitely not the case. For there are parts of Physics, among them the wide region of the phenomena of interference, where the classical theory has proved its validity in every detail, even when subjected to the most delicate measurements; while the Quantum Theory, at least in its present form, is in these respects completely useless. It is not the case that the Quantum Theory cannot be applied, but that, when applied, the results reached do not agree with experience.

The result of this state of affairs is that at the present moment each theory has what may be called its own preserve, where it is safe from attack, while there is also an intermediate region – e.g. that of the phenomena of the dispersion and scattering of light – where the two theories compete with varying fortunes. The two theories are approximately of equal usefulness, so that physicists are guided in the choice of theory by their private predilections – an uncomfortable and, in the long run, an intolerable state of affairs for anyone desirous of reaching the true facts.

To illustrate this curious condition of things I will select a particular example from a very large number collected by workers in the field of theory and of practice. I begin by stating two facts. Let us imagine two fine pencils of rays of violet light, produced by placing an opaque screen with two small holes over against the light which is given out from a point source. The two pencils of rays emerging from the holes can be reflected so that they meet on the surface of a white wall at some distance away. In this case the spot of light which they jointly produce on the wall is not uniformly bright, but is traversed by dark lines. This is the first fact. The second is this – if any metal that is sensitive to light is placed in the path of one of these rays, the metal will continually emit electrons with a velocity independent of the intensity of the light.

Now if the intensity of the source of light is allowed to decrease, then in the first case, according to all the results hitherto obtained, the dark lines remain quite unchanged; it is only the strength of the illumination that decreases. In the other case, however, the velocity of the electrons emitted also remains quite unchanged, and the only change that takes place is that the emission becomes less copious.

Now how do the theories account for these two facts? The first is adequately explained by the classical theory as follows: at every point of the white wall which is simultaneously illuminated by the two pencils of rays, the two rays which meet at this point either strengthen or else weaken each other, according to the relations between their respective wave-lengths. The second fact is equally satisfactorily explained by the Quantum Theory, which maintains that the energy of the rays falls on the sensitive metal, not in a continuous flow, but in an intermittent succession of more or less numerous, equal and indivisible quanta, and that each quantum, as it impinges on the metal, detaches one electron from the mass. On the other hand, all attempts have failed hitherto to explain the lines of interference by the Quantum Theory and the photoelectric effect by the classical theory. For if the energy radiated really travels only in indivisible quanta, then a quantum emitted from the source of light can pass only through one or else the other of the two holes in the opaque screen; while if the light is sufficiently feeble, it is also impossible for two distinct rays to impinge simultaneously on a single point on the white wall; hence interference becomes impossible. In fact the lines invariably disappear completely, as soon as one of the rays is cut off.

On the other hand, if the energy radiated from a point-source of light spreads out uniformly through space, its intensity must necessarily be diminished. Now it is not easy to see how the velocity with which an electron is emitted from the sensitive metal can be equally great whether it is subjected to very powerful or to very weak radiation. Naturally many attempts have been made to get over this difficulty. Perhaps the most obvious way was to assume that the energy of the electron emitted by the metal is not derived from the radiation falling on it, but that it comes from the interior of the metal, so that the effect of the radiation is merely to set it free in the same way as a spark sets free the latent energy of gunpowder. It has, however, not proved possible to demonstrate that there is such a source of energy, or even to make it appear plausible that there should be such a source. Another supposition is that, while the energy of the electrons is derived from the radiation impinging upon them, the electrons themselves are not actually emitted from the metal until this has been subjected to the illumination for a time sufficiently long to allow the energy necessary for a definite velocity to have been accumulated. This process, however, might take minutes or even hours, whereas in fact the phenomenon repeatedly takes place very much sooner. Light is thrown on the profound importance of these difficulties by the fact that in highly influential quarters the suggestion has arisen of sacrificing the validity of the principle of the conservation of energy. This may well be described as a desperate remedy; in this particular instance, in fact, it was soon proved to be untenable by means of experiments.

Hitherto, then, all attempts to understand the laws of the emission of electrons from the standpoint of the classical theory have failed. On the other hand these, and a number of other laws relating to the reciprocal action of radiation and matter, become immediately intelligible and even necessary as soon as we assume that light quanta travel through space in the shape of minute, individual structures and that, when impinging upon matter, they behave like really substantial atoms.

We are compelled, however, to decide in favour of one or the other view; so that the whole problem obviously resolves itself into the question whether the radiant energy emitted from the source of light is divided when it leaves this source, so that one part of it passes through one of the holes in the opaque screen and the remainder through the other, or whether the energy passes in indivisible quanta alternately through each of the two holes. Every theory of quanta must answer this question, and must deal with it in some manner or other; hitherto, however, no physicist has succeeded in giving a satisfactory answer.

It has sometimes been suggested that the difficulties of the Quantum Theory do not so much apply to the propagation of radiation in free space, as to the reciprocal action which takes place between radiation and matter carrying an electric charge. With this opinion I cannot agree. The question set out above confines itself to the propagation of radiation, and there is no reference either to its causes or to its effects.

It might indeed be asked whether we have a right to speak of the energy of free radiation as though it were something actual, since the fact is that all measurements invariably relate to events taking place in material bodies. If we wish to maintain the absolute validity of the energy principle, a standpoint which recent investigation renders particularly plausible, then there can be no doubt that we must assign to every field of radiation a quite definite, and more or less exactly calculable, amount of energy, which is decreased by the absorption of radiation and increased by its emission. The question now is, what is the behaviour of this energy? Once this question is asked, it becomes plain beyond the possibility of doubt that we must make up our minds to admit certain extensions and generalizations of some of the primary assumptions from which we are accustomed to start in theoretical physics, and which hitherto have proved their worth in every field. This becomes necessary in order to find a way out of the difficulty of our dilemma; and it is a result which is sufficiently unsatisfactory to our desire for knowledge. Some consolation can be derived if we see that there is at any rate a possibility of solving the difficulty; consequently I cannot resist the temptation to devote a few words to discussing in what direction it might be possible to find a solution.

The most radical method of avoiding every difficulty would, no doubt, consist in giving up the customary view which holds that radiant energy is localized in some manner or other; i.e. that at every part of a given electromagnetic field, a given amount of energy exists at a given time. If once this assumption is surrendered, the problem ceases to exist, simply because the question whether a light quantum passes through one or the other hole in the opaque screen ceases to have any definite physical meaning. In my opinion, however, this desperate escape from the dilemma goes somewhat too far. For radiant energy as a totality possesses a definite calculable amount; further, the electromagnetic vector-field which is formed by a ray is described in all its optical details, and in the whole of its temporospatial behaviour, by classical electrodynamics, and this description agrees exactly with the facts; finally the energy arises and disappears simultaneously with the field. Consequently it is not easy to avoid the question how the distribution of the energy is affected by the details of the field.

Let us decide to pursue this question as far as possible. Then in order to avoid the alternatives with which we are faced, it might appear expedient to retain the fixed connection between the ray, or rather between the electromagnetic wave on the one hand, and the energy attaching to it on the other, but, while retaining it, to give it a wider and less simple meaning than it has in the classical theory. The latter assumes that every part, however small, of an electromagnetic wave contains a corresponding amount of energy proportional to its magnitude, which is supposed to spread concomitantly with the wave. Now if for this fixed connection we substitute something less rigid, it might then appear that the wave emitted from the source of light divides into any number of parts, in conformity with the classical theory, but that at the same time, in accordance with the Quantum Theory, the energy of the wave is concentrated at certain points. The necessary assumption would be that the energy of the wave is not intimately connected with it in its finest detail. On such an assumption, the phenomena of interference would be explained on the lines that even the weakest wave passes partly through one and partly through the other hole in the opaque screen; while on the other hand the photoelectric effect could be explained on the lines that the wave allows its energy to impinge on the electrons only in integral quanta. Here the difficulty consists in trying to imagine part of a light-wave without the energy appropriate to its magnitude; but though I admit that this is a considerable difficulty, I do not consider it to be essentially greater than that of imagining part of a body without the matter appropriate to its density. Yet we are compelled to make this latter assumption by the fact that matter loses its simple properties if it is subjected to continuous spatial subdivision, since in this case its mass ceases to remain proportional to the space occupied by it, and resolves itself into a number of distinct molecules having a given magnitude. It might well be that the case is closely analogous for electromagnetic energy and the momentum attaching to it.

Hitherto it has been the practice to look for the elementary laws of electromagnetic processes exclusively in the sphere of the infinitely small. Spatially and temporally all electromagnetic fields were divided into infinitely small parts; and their entire behaviour, so far as it appeared subject to laws, was invariably represented by temporo-spatial differential equations. Now in this respect we must radically change our views. For it has been discovered that these simple laws cease to apply after a certain stage in the process of subdivision has been reached, and that beyond this point the increasingly delicate processes make matters more complicated. The spatiotemporal magnitudes of the action become atomic, and we are compelled to assume the existence of elements or atoms of this action. It is indeed a sufficiently striking fact that not a single one of the laws where the universal quantum of action plays a part is expressed by means of a differential equation with a number of continuous variables, but that they all relate to finite times and finite spaces, and deal with such things as definite periods of oscillation, definite orbits, definite transitions, etc. Hence it appears that in order to allow for these facts we must substitute, at least in part, relations between magnitudes at finite distances from each other for those between magnitudes infinitely close to each other. If this is done finite differences take the place of the differential, discontinuity that of continuity, and arithmetic that of analysis; though the substitution admittedly is not carried out radically. A radical substitution is made impossible if only by the claims of the wave theory.

In this direction promising steps have been taken through the development of so-called Quantum Mechanics. This line of investigation has recently produced excellent results in the hands of the Göttingen school of physicists – of Heisenberg, Born and Jordan. Later developments will show how far we can advance towards a solution of the problem along the avenue opened by Quantum Mechanics. Even the choicest mathematical speculations remain in the air so long as they are unsubstantiated by definite facts of experience; and we must hope and trust that the experimental skill of physicists, which in the past has so often definitely decided questions full of doubt and difficulty, will succeed in resolving the difficulties of the present obscure question. In any case there can be no doubt that the parts of the structure of classical Physics, which have had to be discarded as valueless under the pressure of the Quantum Theory, will be supplanted by a sounder and more adequate structure.

To conclude: we have seen that the study of Physics, which a generation ago was one of the oldest and most mature of natural sciences, has today entered upon a period of storm and stress which promises to be the most interesting of all. There can be little doubt that in passing through this period we shall be led, not only to the discovery of new natural phenomena, but also to new insight into the secrets of the theory of knowledge. It may be that in the latter field many surprises await us, and that certain views, eclipsed at the moment, may revive and acquire a new significance. For this reason a careful study of the views and ideas of our great philosophers might prove extremely valuable in this direction.

There have been times when science and philosophy were alien, if not actually antagonistic to each other. These times have passed. Philosophers have realized that they have no right to dictate to scientists their aims and the methods for attaining them; and scientists have learned that the starting-point of their investigations does not lie solely in the perceptions of the senses, and that science cannot exist without some small portion of metaphysics. Modern Physics impresses us particularly with the truth of the old doctrine which teaches that there are realities existing apart from our sense-perceptions, and that there are problems and conflicts where these realities are of greater value for us than the richest treasures of the world of experience.

64

PHANTOM PROBLEMS IN SCIENCE

A Lecture delivered in Göttingen on June 17, 1946.

The world is teeming with problems. Wherever man looks, some new problem crops up to meet his eye – in his home life as well as in his business or professional activity, in the realm of economics as well as in the field of technology, in the arts as well as in science. And some problems are very stubborn; they just refuse to let us in peace. Our agonized thinking of them may sometimes reach such a pitch that our thoughts haunt us throughout the day, and even rob us of sleep at night. And if by lucky chance we succeed in solving a problem, we experience a sense of deliverance, and rejoice over the enrichment of our knowledge. But it is an entirely different story, and an experience annoying as can be, to find after a long time spent in toil and effort, that the problem which has been preying on one's mind is totally incapable of any solution at all – either because there exists no indisputable method to unravel it, or because considered in the cold light of reason, it turns out to be absolutely void of all meaning – in other words, it is a *phantom problem*, and all that mental work and effort was expended on a mere nothing. There are many such phantom problems – in my opinion, far more than one would ordinarily suspect – even in the realm of science.

There is no better safeguard against such unpleasant experiences than to ascertain in each instance, and at the very outset, whether the problem under consideration is a genuine

66

or meaningful one, and whether a solution for it is to be expected. In view of this situation I will cite and examine a number of problems, in order to see whether they happen to be mere phantom problems. By doing so, I may be able to render a genuinely useful service to some of you. My selection of these problems to be exhibited as specimens is not based on any systematic viewpoint, and even less can it lay a claim to completeness in any respect. Most of them are taken from the realm of science, because this is the field in which the relevant factors are the most clearly discernible. However, this consideration will not deter me from touching upon other fields, too, whenever I can reasonably surmise that the subject holds an interest for you.

I

In order to decide whether or not a given problem is truly meaningful, we must first of all examine closely the assumptions contained in its wording. In many instances, these alone will immediately reveal the problem under consideration to be a phantom problem. The matter is simplest when an error is lurking in the assumptions. In this case, of course, it is immaterial whether the erroneous assumption was introduced deliberately or has just escaped detection. A lucid example is the famous problem of perpetual motion, i.e. the problem of devising a periodically functioning apparatus which will perform mechanical work perpetually without any other change in nature. Since the existence of such an apparatus would contradict the principle of the conservation of energy, such an apparatus cannot possibly occur in nature, so that this problem is a phantom problem. Of course, one may raise the following argument: "The principle of the conservation of energy, after all, is an experimental law. Accordingly, although today it is considered to be universal and all-embracing, its validity may one day have to be restricted – and in fact, such a curtailment of its universal applicability has been sometimes suspected in nuclear physics – and the problem of perpetual motion would then suddenly become genuine. Its meaninglessness is, therefore, by no means absolute."

This counter-argument may actually acquire practical significance, as is demonstrated especially clearly by the example of a no less well-known problem in chemistry: The ancient problem of changing base metal, for instance, mercury, into gold. Originally, prior to the birth of a scientific chemistry, this problem was considered to be pregnant with portentous meaning, and many a learned – and unlearned – mind was zealously occupied with it. But later, as the theory of chemical elements was developed and became universally accepted, the transmutation of metals turned into a phantom problem. In recent times, since the discovery of artificial radioactivity, the situation has again been reversed. The fact is that today it no longer seems to be fundamentally impossible to discover a process for removing a proton from the nucleus of the mercury atom and an electron from its shell. This operation would change the mercury atom into a gold atom. Therefore, at the present stage of science, the ancient quest of the alchemists no longer belongs to the class of phantom problems.

However, these examples must by no means be construed as indicating that the meaninglessness of a phantom problem is never absolute, but simply dependent on whether or not a certain theory is accepted as valid. There are also many phantom problems which are indubitably doomed to remain such forever. One of these, for instance, is the problem which used to keep many a great physicist busy for many years: The study of the mechanical properties of the luminiferous ether. The meaninglessness of this problem follows from its basic premise, which postulates that light vibrations are of a mechanical nature. This premise is erroneous, and must so remain forever.

Here is another example, taken from the field of physiol-

ogy: It is a well known fact that the convex lens of the human
eye projects an inverted image on the retina. When we see
a tower, its image appears on the retina with the top of the
tower pointing downward. When this phenomenon was established, a number of scientists tried to detect in the human
organ of sight that particular mechanism which supposedly
re-inverts the image on the retina. This is a phantom problem, and never can be anything else, for it is based on an
erroneous premise, for which there can be no possible proof
– namely, that in the organ of vision the image of an object
must be upright rather than inverted.

Far more difficult than those cases in which, as in the examples just cited, the assumptions are mistaken, are problems
whose presuppositions contain no error, but are so vaguely
worded that they must remain phantom ones because they
are inadequately formulated. Yet, it so happens that it is
just such cases with which we shall be chiefly preoccupied.

My first example is a phantom problem, for the triviality
of which I beg your forgiveness. The room in which we now
sit, has two side walls, a right-hand one and a left-hand one.
To you, this is the right side, to me, sitting facing you, that
is the right side. The problem is: Which side is in reality the
right-hand one? I know that this question sounds ridiculous,
yet I dare call it typical of an entire host of problems which
have been, and in part still are, the subject of earnest and
learned debates, except that the situation is not always quite
so clear. It demonstrates, right at the very outset, what great
caution must be exercised in using the word, *real.* In many
instances, the word has any sense at all only when the speaker
first defines clearly the point of view on which his considerations are based. Otherwise, the words, *real* or *reality* are
often empty and misleading.

Another example: I see a star shining in the sky. What is
real in it? Is it the glowing substance, of which it is composed,
or is it the sensation of light in my eyes? The question is

meaningless so long as I do not state whether I am assuming a realistic or a positivistic point of view.

Still another example, this one from the realm of modern physics: When the behavior of a moving electron is studied through an electron microscope, the electron appears as a particle following a definite course. But when the electron is made to pass through a crystal, the image projected on the screen shows every characteristic of a refracted light wave. The question, whether the electron is in reality a particle, occupying a certain position in space at a certain time, or a wave, filling all of infinite space, will therefore constitute a phantom problem so long as we fail to stipulate which of the two viewpoints is applied in the study of the behavior of the electron.

The famous controversy between Newton's emission theory and Huygens' wave theory of light is also a phantom problem of science. For every decision for or against either of these two opposing theories will be a completely arbitrary one, depending on whether one accepts the point of view of the quantum theory or that of the classical theory.

II

In every one of the cases cited till now, we encountered a rather simple, easily appreciable situation. Now let us proceed to the consideration of a problem which has always been regarded as of central importance because of its meaning to human life – the famous body-mind problem. In this case, first of all we must try to ascertain the meaning of our problem. For there are philosophers who claim that mental processes need not be accompanied by physical processes at all, but can take place totally independently from the latter. If this view is right, mental processes are subject to entirely different laws than those applying to physical processes. If so, then, the body-mind problems splits into two separate problems – the body problem and the mind problem – thus losing its meaning, and degenerating into a phantom problem. With this finding, the case may be considered as good as closed, and we need only concern ourselves with the reciprocal interaction of mental and physical processes. Experience shows that they are very closely influenced by each other. For instance, somebody asks me a question. His question is introduced by a physical process, the propagation of the sound waves of the spoken words which, emitted by him, hit my ears and are transmitted to my brain through the sensory nerve paths. They then cause mental processes to take place in my brain, namely, a reflection on the meaning of the words perceived, followed by a decision as to the content of the answer to be given. Then another physical process operates my motor nerves and my larynx, to transmit the answer to the questioner by means of the physical process of propagating sound waves through the air.

Now then, what is the nature of the interrelation of physical and mental processes? Are mental processes caused by physical ones? And if so, according to what laws? How can something material act on something immaterial, and *vice versa*? All these questions are difficult to answer. If we assume the existence of a causal interaction, a cause-and-effect relationship, between physical and mental processes, a continued, unrestricted applicability of the principle of the conservation of energy appears to be an indispensable premise. For one will not be disposed to sacrifice this universal foundation of exact science. But in that case, there would have to exist a numerically definite mechanical equivalent of psychic processes, as there is a definite equivalent of heat in thermodynamics, and there would be absolutely no method for measuring such a constant. For this reason, a solution has been attempted on the basis of the hypothesis that the mental forces contribute no perceptible energy to the physical processes, but act merely to liberate the latter, as a gentle breeze will start something that will grow into a mighty avalanche, or a tiny spark will blow up a huge powder magazine. However, this hypothesis does not solve the difficulty completely. Because in every case known to us, while the amount of energy expanded in liberating a process is very small in comparison with the energy released, yet it does exist, even though it may have just a microscopic magnitude. Even the very gentlest breeze and very tiniest spark possess an energy above zero – and this is what matters here.

However, it is well known that there are some forces which produce a perceptible effect without any expenditure whatever of energy. These are what we may call "guiding forces," such as, for instance, the resistance due to the rigidity of railroad rails which forces the wheels of a train to follow a pre-detennined curve, without any expenditure of energy. An attempt might be made to ascribe a similar role to the mental forces in the guiding of physical processes along predetermined paths in the human brain. But this, too, involves grave and insurmountable difficulties. For the modern science of brain physiology is based on the very premise that it is possible to achieve a satisfactory understanding of the laws of biological processes without postulating the intervention of any particular mental force. Such a hypothesis avoids also the theory of parallelism which, in contrast to the theory of interaction, assumes that mental and physical processes must, necessarily, run side by side, each according to its own laws, without interfering with each other. Of course, it still remains incomprehensible just how this reciprocal interdependence of two such fundamentally different occurrences is to be conceived, and whether it perhaps requires the assumption of some form of pre-established harmony. In this respect, the theory of parallelism, too, is hardly satisfactory.

And now, in order to get to the bottom of the matter, let us ask ourselves this question of basic significance: Just what do we know about mental processes? In what circumstances and in what sense may we speak of mental processes? Let us consider first where we come across mental processes in this world. We must take it for granted that members of the higher animal kingdom as well as human beings have emotions and sensations, But as we descend to the lower animals – where is the borderline where sensation ceases to exist? Has a worm any sensation of pain as it is crushed under our feet? And may plants be considered capable of some kind of sensation? There are botanists who are disposed to answer this question affirmatively. But such a theory can never be put to the test, let alone proved, and the wisest course seems to be not to venture any opinion in this regard. Along the entire ladder of evolution, from the lowest order of life up to Man, there is no point at which one can establish a discontinuity in the nature of mental processes.

It is nevertheless possible to specify a quite definite borderline, of decisive importance for all that follows. This is the borderline between the mental processes within other individuals and the mental processes within one's own Ego. For everybody experiences his own emotions and sensations directly. They just simply exist for him. But we do not experience directly the sensations of any other individuals, however certain their existence may be, and we can only infer them in analogy to our own sensations. To be sure, there are physicians who solemnly claim to be able to perceive the emotions and moods of their patients no less clearly than the latter themselves. But such a claim can never be proved indisputably. Its questionability becomes most striking if we think of certain specific instances. Even the most sensitive dentist cannot feel the piercing pains which his patient at times has to suffer under his treatment. He can ascertain them only indirectly, on the basis of the moans or squirming of the patient. Or, to speak of a more pleasant situation, such as for instance a banquet, however clearly one may sense the pleasure of one's neighbor over the taste of his favorite wine, it is something quite different from tasting it on one's own tongue. What *you* feel, think, want, only *you* can know as first-hand information. Other people can conclude it only indirectly, from your words, conduct, actions and mannerisms. When such physical manifestations are entirely absent, they have no basis whatever to enable them to know your momentary mental state.

This contrast between first-hand, or direct, and second-hand, or indirect, experience is a fundamental one. Since our primary aim is to gain direct, first-hand experience, we shall now discuss the interrelation of our mental and physical states.

First of all, we find that we may speak of conscious states only. To be sure, many processes, perhaps even the most decisive ones, must be taking place in the subconscious mind. But these are beyond the reach of scientific analysis. For there exists no science of the unconscious, or subconscious, mind. It would be a contradiction in terms, a self-contradiction. One does not know that which is subconscious. Therefore, all problems concerning the subconscious are phantom problems.

Let us therefore take a simple conscious process involving body and mind. I prick my hand with a needle, and feel a sensation of pain. The wound made by the pin is the physical element, the sensation of pain is the mental element of the process. The wound is seen, the pain is felt. Is there, then, an indisputable method of throwing light upon the interrelation of the two elements of this process? It is easy to realize that this is absolutely impossible. For there is nothing here upon which light is to be thrown. The visual perception of the wound and the feeling of the pain are elementary facts of experience, but they are as different in nature as knowledge and feeling. Therefore, the question as to their essential interrelation represents no meaningful problem – it is just a phantom problem.

It is obvious that the two occurrences, the pinprick and

the sensation of pain, can be examined and analyzed most thoroughly, in every detail. But such an analysis calls for two different methods, which mutually preclude each other. Each of the two corresponds to one of two different viewpoints. In the following I will refer to them, respectively, as the *psychological* and the *physiological* viewpoints. Observation based on the psychological viewpoint is rooted in self-consciousness; therefore, it is applicable directly only to the analysis of one's own mental processes. On the other hand, observation based on the physiological viewpoint is directed at the processes in the external world; therefore, its direct scope is limited to physical processes. These two viewpoints are incompatible. The adoption of one when the other one is called for, always leads to confusion. We cannot judge our mental processes directly from the physiological viewpoint any more than we can examine a physical process from, the psychological viewpoint. This state of affairs makes the body-mind problem appear in a different light. For the examination of psychosomatic processes will yield entirely different results, according as the psychological or the physiological viewpoint is taken as the basis of observation. The psychological viewpoint will permit us to gain first-hand knowledge solely and exclusively of something that relates to our mental processes. The physiological viewpoint will produce first-hand information about physical processes only. It is therefore impossible to gain first-hand information about both physical and mental processes from any single viewpoint; and since in order to reach a clear conclusion, we must adhere to a given viewpoint, which automatically excludes the other, the search for the interrelation of physical and mental processes loses its meaning. In this case, there exist only physical processes *or* mental processes, but never processes which are physical *and* mental.

Therefore, it will do no harm to say that the physical and the mental are in no way different from each other. They are the selfsame processes, only viewed from two diametrically opposite directions. This statement is the answer to the riddle, which has been inseparable from the theory of parallelism, namely, how one is to conceive the fact that two types of processes so different from each other as the physical and the mental, are so closely interlinked. The link has now been disclosed. At the same time, the body-mind problem has been recognized as another phantom problem.

III

The cases heretofore discussed have dealt only with knowledge, and feeling. Physical states and processes are known, mental states and processes are felt. The situation is quite different, and more complicated, when cognition and feeling are joined by volition. For in that case we are confronted by the ancient dilemma of freedom of the will *versus* the law of causality. This problem holds a certain significance for ethics, too, and its discussion will be our immediate next step.

Is the will free, or is it causally determined? In order to be able to answer this question, first of all we must examine the methods which can be utilized for the study of the laws and regularity of volitional processes.

In this connection an important point must first be observed: In order to gain a correct insight into the regular course of a process, one must take every precaution lest the process be influenced by the method of observation used. Thus, for instance, when trying to ascertain the temperature of a body, we must not use any thermometer, the introduction of which would cause a change in the temperature under examination; similarly, in the microscopic observation of the processes taking place in a living cell, we must not employ illumination which might interfere with the normal course of those processes. All that holds true for physical and biological processes, applies naturally to the same extent to mental states and processes, too. It is one of the most

elementary principles of experimental psychology that an observation may produce a totally false finding if the subject
knows, or even suspects that he is being observed. For this
reason, under certain circumstances, the observation itself will
constitute a serious source of error.

Applying the above principle to the problem now before
us, the most basic and elementary requirement which a scientifically perfect observation of the regular course of a volitional impulse must fulfill, is that it should not affect or
influence that impulse. The automatic consequence of this
requirement is that the choice of the acceptable viewpoint
of the observation must, necessarily, be restricted. Namely,
since the observation itself is no less a mental process than
the volitional impulse which is to be observed, the observation may, under certain circumstances, influence the course of
the volitional impulse, and thus distort the final finding. The
only time when there is no reason to fear such interference
is when you observe the will of another person without his
knowledge, or when another person observes your will without your knowledge. On the other hand, this source of error
will always be operative whenever you attempt to observe
your own will. For in that case the mental process of the
observation coincides in your unified self-consciousness with
the mental process of the volitional impulse. Therefore, it is
inadmissible to observe one's own will from the viewpoint of
one's own Ego – and I am referring to the present as well as
the future act of will, for the latter is co-determined by the
present will, too. On the other hand, there is nothing to preclude a scientific observation of a volitional impulse of one's
own past Ego. For past mental processes are not affected by
a later analysis. In order to express this situation, I shall
from now on make a distinction between an external and an
internal viewpoint of observation. The external viewpoint is
the one which permits the volitional processes to be observed
without being disturbed, affected, or interfered with, by the

observation. This viewpoint is adopted when observing the volitional processes of others, as well as when observing the *past* volitional processes of one's own Ego. The internal viewpoint is the one, from which the volitional processes cannot be observed without being thereby disturbed. This viewpoint is, adopted when observing the present and future volitional processes of one's own Ego. The external viewpoint is suitable for a scientific examination of the laws governing volitional processes; the internal viewpoint is not admissible for this purpose. It is self-evident that these two view-points mutually preclude each other, and that it is senseless to apply both of them simultaneously.

Now then, if we adopt the external viewpoint – the only one admissible – as the basis of our scientific observation of volitional processes, everyday experience tells us that in our daily dealings with others we always presuppose certain motives, in other words, a causal determinism, in whatever they say and do, for otherwise their behavior would be inaccountable, and any orderly contact with them would be impossible. The same principle applies to scientific research, too. If a historian wanted to ascribe the decision of Julius Caesar to cross the Rubicon not to his political deliberations and to his innate temperament, but to free will, his view would be tantamount to a renunciation of scientific understanding. Therefore, we will have to conclude that from the external viewpoint of observation the will is to be assumed as causally determined.

The state of affairs is quite different as regards the internal viewpoint. As we have seen, the scientific method of observation fails to work here. On the other hand, however, this viewpoint opens up another source of information: Self-consciousness, which tells us immediately that we are able at any time to give any desired turn to our will as we can to our thoughts, whether as a result of mature deliberation, discretion, or even sheer whim. In this connection, it must be observed that this is by no means a matter of an overt volitional act, which is often impeded by external circumstances, but solely giving the will an intended direction. In this domain we have absolute supreme command. Just think of the unspoken mental reservations which we are able to make with every word we speak. This is a real freedom, experienced at first hand, not a make-believe freedom, as it is claimed by many people who are unable to keep distinct the two opposite viewpoints. Of course, he who seeks to know the "real" freedom of will without reference to the view-point adopted, proceeds no differently than does the one who asks without further specification which side of this room is "really" the right side. On the present analysis, neither does the freedom of the will rest, as some have supposed, on a certain lack of intelligence. The degree of intelligence is of absolutely no significance here. Even the most intelligent person is no more capable of observing himself from the outside than is even the fastest runner of passing himself.

In summary, we can therefore say: Observed from without, the will is causally determined. Observed from within, it is free. This finding takes care of the problem of the freedom of the will. This problem came into being because people were not careful enough to specify explicitly the viewpoint of the observation, and to adhere to it consistently. This is a typical example of a phantom problem. Even though this truth is still being disputed time and again, there is no doubt in my mind that it is but a question of time before it will gain universal recognition.

IV

There are many more examples that can be cited to illustrate what grave consequences may be entailed by an improper confusion of two opposite viewpoints. Let us consider one more particularly frequent case, the confusion of the scientific viewpoint with the religious viewpoint. Even though science and religion, in their ultimate effects, are headed for the same goal, the recognition of an omnipotent intellect ruling the universe, yet they are basically different both in their starting points and methods. And in order to attain fruitful results, one must be careful to study a given problem from the suitable viewpoint, which must then be consistently followed. Unfortunately, up to this time, this requirement has often been totally disregarded. In fact, inquiries often switch abruptly from one perspective to the other. This error is being committed by both sides; in other words, one often finds the scientific viewpoint improperly applied to the treatment of ethico-religious questions, and also considerations of religious character dragged into purely scientific problems. The first case is illustrated by the above discussion of the consciousness of freedom, which recent attempts have endeavored to reduce to the breakdown of the law of causality in modern physics, although it has nothing in the least tp do with the law of causality. The repeated efforts to establish scientific grounds for the existence and personality of God stand on the same level. The other side of the picture is exemplified by the violent battle of the Church in past ages against the Gopernican view of the universe, or in recent times by the campaign against the physical theory of relativity, on the basis of emotional attitudes and political arguments, which have nothing to do with science.

In this matter, a fundamental and serious dilemma must, however, be faced. If so many instances indicate that great and important questions are revealed to be phantom problems on careful analysis, that in fact the very word, *real*, often has a meaning which varies with the standpoint adopted – does not scientific knowledge reduce to a plain relativism? Is there, then, no absolutely valid view, no absolute reality, independent of any special perspective?

It would be unfortunate indeed if this were so. No – there must exist, in science, too, absolutely correct and final maxims, just as there are absolute values in ethics. Moreover, and this is the main thing, these very propositions, maxims and values are the most important and worth while goals of every endeavor. In the realm of exact science, there are the values of the *absolute constants*, such as the elementary quantum of electricity, or the elementary quantum of action, and many others. These constants always prove to be the same, regardless of the method used for measuring them. The endeavor to discover them and to trace all physical and chemical processes back to them, is the very thing that may be called the ultimate goal of scientific research and study.

Nor is the situation different in the world of religion and ethics. To be sure, there, too, a considerable role is often played by the viewpoint that is adopted as a consequence of the special conditions involved in a given problem. Thus, the moral standard of truthfulness often appears to be loosened and weakened in a regrettable manner, I want to disregard here completely the conventional lies to which people resort for the sake of social amenities. But truthfulness, this noblest of all human virtues, is authoritative even here over a well-defined domain, within which its moral commandment acquires an absolute meaning, independent of all specific viewpoints. This is probity to one's own self, before one's own conscience. Under no circumstances can there be in this domain the slightest moral compromise, the slightest moral justification for the smallest deviation. He who violates this commandment, perhaps in the endeavor to gain some momentary worldly advantage, by deliberately and knowingly shutting his eyes to the proper evaluation of the true situation, like a spendthrift who thoughtlessly squanders away his wealth, must inevitably suffer, sooner or later, the grave consequences of his foolhardiness.

These absolute values in science and ethics are the ones whose pursuit constitutes the true task of every intellectually alert and active human being, a task which confronts all men again and again, in one form or another. This task is never finished – a fact guaranteed by the circumstance that genuine problems, even though sometimes accompanied by phantom problems, constantly appear in ceaseless variety and constantly set new tasks for active human beings. For it is work which is the favorable wind that makes the ship of human life sail the high seas, and as for the evaluation of the worth of this work, there is an infallible, time-honored measure, a phrase which pronounces the final, authoritative judgment for all times: *By their fruits ye shall know them!*

THE MEANING AND LIMITS OF EXACT SCIENCE

A Lecture delivered in November 1947.

Exact Science – what wealth of connotation these two words have! They conjure up a vision of a lofty structure, of imperishable slabs of stone firmly joined together, treasure-house of all wisdom, symbol and promise of the coveted goal for a human race thirsting for knowledge, longing for the fi-nal revelation of truth. And since knowledge always means power, too, with every new insight that Man gains into the forces at work in Nature, he always opens up also a new gate-way to an ultimate mastery over them, to the possibility of harnessing these natural forces and making them obey his every command.

But this is not all – nor even the most important part of it. Man wants not only knowledge and power. He wants also a standard, a measure of his actions, a criterion of what is valuable and what is worthless. He wants an ideology and philosophy of life, to assure him of the greatest good on earth – peace of mind. And if religion fails to satisfy his longing, he will seek a substitute in exact science. I refer here merely to the endeavors of Monism, founded by outstanding schol-ars, philosophers and natural scientists, a school of thought which commanded high respect only as recently as a short generation ago.

Yet, in these our days hardly a word is being heard about

the Monists, although the structure of their ideology was unquestionably erected to endure for a long time to come, and it started out on its career with high hopes and great promises. There must be something wrong somewhere! And in fact, if we take a closer look and scrutinize the edifice of exact science more intently, we must very soon become aware of the fact that it has a dangerously weak point – namely, its very foundation. Its foundation is not braced, reinforced properly, in every direction, so as to enable it to withstand external strains and stresses. In other words, exact science is not built on any principle of such universal validity, and at the same time of such portentous meaning, as to be fit to support the edifice properly. To be sure, exact science relies everywhere on exact measurements and figures, and is therefore fully entitled to bear its proud name, for the laws of logic and mathematics must undoubtedly be regarded as reliable. But even the keenest logic and the most exact mathematical calculation cannot produce a single fruitful result in the absence of a premise of unerring accuracy. Nothing can be gained from nothing.

No phrase has ever engendered more misunderstanding and confusion in the world of scholars than the expression, "*Science without Presuppositions.*" It was coined originally by Theodor Mommsen, and was meant to express that scientific analysis and research must steer clear of every preconceived opinion. But it could not be, nor was it, intended to mean that scientific research needs no presuppositions at all. Scientific thought must link itself to something, and the big question is, *where.* This question has occupied the minds of the most profound thinkers of all epochs and all nations, since time immemorial, from Thales to Hegel, setting in motion all forces of man's imagination and logic. But it has been demonstrated again and again that a final, conclusive answer cannot be found. Perhaps the most impressive proof of this negative finding is that until now all attempts have failed to discover a world view uniformly acceptable, in its general features at least, by all minds capable of judgment. The only conclusion which this fact permits, according to every dictate of reason, is that it is absolutely impossible to place exact science in an *a priori* manner on a universal foundation possessing a fixed and inclusive content.

Thus, at the very outset of our quest for the meaning of exact science, we are confronted by an obstacle which must be a disappointment to everybody who is seriously striving for knowledge. In fact, this obstacle has driven many a critically disposed thinker to join the ranks of the skeptics. And a no less regrettable fact is that there are perhaps just as many, or even more, individuals of the opposite disposition whom the fear of falling victims to skepticism – an ideology which they consider intolerable – drives to look for salvation to prophets of creeds like, for instance, anthroposophy. Such prophets appear on the scene in all epochs, not excepting our own, with their brand new message of salvation, and they often succeed, with an amazing rapidity, in gathering a following of enthusiastic disciples, eventually to make their exit from the stage and to sink back into the all-engulfing abyss of oblivion.

Is there a way out of this fatal dilemma? And where can it be found? This is the first question to claim our attention. I shall attempt to show that there is a positive answer to it, and that this answer will cast a light both on the meaning and limits of exact science. I submit to the judgment of each of you the validity of my proposed resolution of the problem.

I

If we seek a foundation for the edifice of exact science which is capable of withstanding every criticism, we must first of all tone down our demands considerably. We must not expect to succeed at a stroke, by one single lucky idea, in hitting on an axiom of universal validity, to permit us to develop, with exact methods, a complete scientific structure.

We must be satisfied initially to discover some form of truth which no skepticism can attack. In other words, we must set our sights not on what we would like to know, but first on what we do know with certainty,

Now then, among all the facts that we do know and can report to each other, which is the one that is absolutely the most certain, the one that is not open even to the most minute doubt? This question admits of but one answer: "That which we experience with our own body." And since exact science deals with the exploration of the outside world, we may immediately go on to say; "They are the impressions we receive in life from the outside world directly through our sense organs, the eyes, ears, etc." If we see, hear or touch something, it is clearly a given fact which no skeptic can endanger.

To be sure, we speak also of illusions, but never with the intention of implying that the sense perceptions involved are incorrect or even questionable. For instance, when a person happens to be deceived by a mirage, the fault lies not with his perception of the visual image, which is actually present, but in his inferences which draw false conclusions from the given sensory data. The sensory impression is always a given fact, and therefore incontestable. What conclusions the individual attaches to it, is another story, which need not concern us for the time being. Therefore, the content of the sensory impressions is the most suitable and only unassailable foundation on which to build the structure of exact science.

If we call the sum total of sensory impressions "the sense world," we may state briefly that exact science issues from the experienced sense world. The sense world is that which, so to speak, furnishes science with the raw material for its labors.

However, this seems to be a very meager result. For the content of the sense world is, in any case, only something of a subjective character. Every individual has his own senses, and in general, the senses of one individual are quite different from those of another, whereas the aim of exact science is to achieve objective, universally valid knowledge. It may seem, therefore, that in adopting our present approach we have been following the wrong track.

But we must not jump to conclusions. For it will become manifest that considerable progress can be made along the line of advance now open to us. Considered as a whole, the matter reduces itself to the fact that we human beings have no direct access to the knowledge conveyed to us by exact science, but must acquire it one by one, step by step, at the cost of painstaking labors of years and centuries.

Now, if we examine the content of our sense world, it obviously falls apart into as many separate fields as we have sense organs – there is a field corresponding to sight, another to hearing, and still others corresponding to the senses of touch, smell, taste and heat. These fields are totally different from each other, and have initially nothing in common. There is no immediate, direct bridge between the perception of colors and the perception of sounds. An affinity, such as may be assumed by many art lovers to exist between a certain shade of color and a certain musical pitch, is not directly given, but is the creation, stimulated by personal experiences, of our reflective power of imagination.

Since exact science deals with measurable magnitudes, it is concerned primarily with those sensory impressions which admit of quantitative data – in other words, the world of sight, the world of hearing, and the world of touch. These fields supply science with its raw material for study and research, and science goes to work on it with the tools of a logically, mathematically and philosophically disciplined reasoning.

II

What, then, is the meaning of this work of science? Briefly put, it consists in the task of introducing order and regularity into the wealth of heterogeneous experiences conveyed by the various fields of the sense world. Under closer examination, this task proves to be fully consistent with the task which we are habitually peforming in our lives ever since our earliest infancy, in order to find our way and place in our environment. This is a task which has kept man busy ever since he first began to think at all in order to be able to hold his own in the struggle for existence. Scientific reasoning does not differ from ordinary everyday thinking in kind, but merely in degree of refinement and accuracy, more or less as the performance of the microscope differs from that of the naked eye. The truth of this statement, and that it must necessarily be so, is evident from the very fact that there is only one kind of logic, and, therefore, even scientific logic cannot deduce anything else from given presuppositions than can the ordinary logic of untrained common sense.

We shall therefore obtain an intuitively clear understanding of the results which science achieves through its labors, if we take our point of departure from the experiences known and familiar to us from daily life. If we review our own personal, individual development, and consider the point which our world view has giadually reached in the course of the years, we can say that we are trying to use the facts of experience as the foundation for a unified, comprehensive and practically serviceable picture of the world in which we live; that we conceive the outside world as filled with objects which act on our various sense organs, thus producing our different sensory impressions.

However, since this practical world picture which every human being carries within himself is not a directly given notion, but an idea elaborated gradually on the basis of facts of experience, it is possessed of no final character. It is changed and adjusted by every new datum of experience, from infancy to adulthood, first at a quicker, then at a slower pace. The same principle applies to the scientific world picture. The scientific world picture or the so-called phenomenological world is also not final and constant, but is in a process of constant change and improvement. It differs from the practical world picture of daily life not in kind, but in its finer structure. It is to the world picture of daily life approximately as the world picture of the adult human being is to the world picture of the human child. Therefore the best start toward a correct understanding of the scientific world picture will be to investigate the most primitive world picture, the naive world picture of the child.

Let us, therefore, try to place ourselves, as best we can, in the child's mind and world of ideas. As soon as the child begins to think, he begins to form his world picture. For this purpose, he directs his attention toward the impressions which he receives through his sense organs. He tries to classify them, and in this endeavor he makes all kinds of discoveries, such as, for instance, that there is a certain orderly interrelation between the inherently different impressions conveyed by the senses of sight, touch and hearing. If you give the child a toy, let us say, a rattle, he will find that the tactile sensation is always accompanied by a corresponding visual sensation, and as he moves the rattle back and forth, he also perceives a certain regular auditory sensation.

While in this instance the different mutually independent sense worlds appear to be interlocking to a certain degree, on other occasions the child will make an observation which he will find to be no less remarkable – that certain impressions which are completely indistinguishable from one another and have their origin in a common sense world, may nevertheless be of a totally different character. Thus, for instance, the child may look at a round electric light and observe that it looks just like the full moon. The light sensation may be exactly the same. Yet, the child finds a great difference, for he can touch the lamp, but not the moon; he can pass his hand around the lamp, but not around the moon.

What, then, does the child think as he makes these discoveries? First of all, he wonders. This feeling of wonderment is the source and inexhaustible fountain-head of his desire for knowledge. It drives the child irresistibly on to solve the mystery, and if in his attempt he encounters a causal relationship, he will not tire of repeating the same experiment ten times, a hundred times, in order to taste the thrill of discovery over and over again. Thus, by a process of incessant labor from day to day, the child eventually develops his world picture, to the degree needed by him in practical life.

The more the child matures, and the more complete his world picture becomes, the less frequently he finds reason to wonder. And when he has grown up, and his world picture has solidified and taken on a certain form, he accepts this picture as a matter of course and ceases to wonder. Is this because the adult has fully fathomed the correlations and the necessity of the structure of his world picture? Nothing could be more erroneous than this idea. No! – The reason why the adult no longer wonders is not because he has solved the riddle of life, but because he has grown accustomed to the laws governing his world picture, But the problem of why these particular laws and no others hold, remains for him just as amazing and inexplicable as for the child. He who does not comprehend this situation, misconstrues its profound significance, and he who has reached the stage where he no longer wonders about anything, merely demonstrates that he has lost the art of reflective reasoning.

Rightly viewed, the real marvel is that we encounter natural laws at all which are the same for men of all races and nations. This is a fact which is by no means a matter of course. And the subsequent marvel is that for the most part these laws have a scope which could not have been anticipated in advance.

Thus, the element of the wondrous in the structure of the world picture increases with the discovery of every new law. This holds true even of scientific research and inquiry in our own day, which continually produce something new. Just think of the mysteries of the cosmic rays, or the mysterious hormones, or the remarkable revelations of the electron microscope. To the research scientist, no less than to the child, it is always a gratifying experience and an added stimulus to encounter a new wonder, and he will labor industriously to solve the riddle by repeating the same experiments with his refined instruments just as the child does with his primitive rattle.

However, let us not leap too far ahead, but proceed in an orderly fashion. First, let us investigate in what respect the structure of the child's world differs from the sense world as originally given. The first fact to claim our attention is that sensations, the sole and exclusive constituents of the original world picture, have been driven appreciably into the background. The dominant elements of this world picture are not sensations, but the objects which produce them. The toy is the dominant element, and the tactile, visual and auditory sensations are merely secondary consequences. But we would not do full justice to the state of affairs were we to say simply that this world picture is nothing but a synthesis of different sensory impressions achieved with the help of the unifying concept of *thing*. For, conversely, a single undifferentiated sense experience may correspond to several different objects. An example of this possibility is the previously mentioned case of an illuminated surface which produces in us a definite sensory impression, and yet is sometimes attributed to an electric light while at other times to the full moon. This is a case of a single undifferentiated sensation which corresponds to two different objects. The contrast, therefore, lies deeper, and can be characterized exhaustively only by introducing the concept of an objectively valid regularity. The sensations produced by objects are private, and vary from one individual to another. But the world picture, the world of objects, is the

92

same for all human beings, and we may say that the transition from the sense world to the world picture amounts to a replacement of a disordered subjective manifold by a constant objective order, of chance by law, and of variable appearance by stable substance.

The world of objects, in contrast to the sense world, is therefore called the *real world*. Yet, one must be careful when using the word, *real*. It must be taken here in a qualified sense only. For this word has the connotation of something absolutely stable, permanent, immutable, whereas the objects of the child's world picture could not rightly be claimed to be immutable. The toy is not immutable, it may break or burn. The electric lamp can be smashed to smithereens. This precludes their being called *real* in the sense just mentioned.

This sounds both self-evident and trivial. But we must bear in mind that in the case of the scientific world picture, where as we have seen, the situation is quite analogous, this state of affairs was by no means found to be self-evident. For just as to the child the toy is the true reality, so for detades and centuries the atoms were taken by science to constitute the true reality in natural processes. The atoms were considered to be that which remains immutable when an object is smashed or burned, thus representing permanency in the midst of all change – until one day, to everybody's astonishment, it was found that even atoms could change. Therefore, whenever in the sequel we refer to the "real world," we shall be using the word *real* primarily in a qualified, naive sense, adjusted to the particular character of the dominant world picture, and we must constantly bear in mind that a change in the world picture may go hand in hand, simultaneously, with that which people call "real."

Every world picture is characterized by the real elements, of which it is composed. The real world of exact science, the scientific world picture, evolved from the real world of practical life. But even this world picture is not final, but changes all the time, step by step, with every advance of inquiry.

Such a stage of development is represented by that scientific world picture which today we are accustomed to call "classical." Its real elements, and hence its characteristic feature, were the chemical atoms. In our own day, scientific research, fructified by the theory of relativity and the quantum theory, stands at the threshold of a higher stage of development, ready to mould a new world picture for itself. The real elements of this coming world picture are no longer the chemical atoms, but electrons and protons, whose mutual interactions are governed by the velocity of light and by the elementary quantum qf action. From today's point of view, therefore, we must regard the realism of the classical world picture as naive. But nobody can tell whether some day in the future the same words will not be used in referring to our modern world picture, too.

III

But what is the meaning of this constant shift in what we call "real"? Is it not utterly unsatisfactory to all men who seek definite scientific insight?

The answer to this question must be, first of all, that our immediate concern is not whether or not the situation is satisfactory, but what its essentials are. But the pursuit of this question will lead to a discovery which we must regard as the greatest of all the wonders previously mentioned. First of all, it must be noted that the continual displacement of one world picture by another is dictated by no human whim or fad, but by an irresistible force. Such a change becomes inevitable whenever scientific inquiry hits upon a new fact in nature for which the currently accepted world picture cannot account. To cite a concrete example, such a fact is the velocity of light in empty space, and another is the part played by the elementary quantum of action in the regular occurrence of all atomic processes. These two facts, and many more, could not be incorporated in the classical world picture, and consequently, the classical world picture had to yield its place to a new world picture.

This in itself is enough to make one wonder. But the circumstance which calls for ever greater wonderment, because it is not self-evidently a matter of course by any means, is that the new world picture does not wipe out the old one, but permits it to stand in its entirety, and merely adds a special condition for it. This special condition evolves a certain limitation, but because of this very fact it simplifies the world picture considerably. In fact, the laws of classical mechanics continue to hold satisfactorily for all the processes in which the velocity of light may be considered to be infinitely great, and the quantum of action to be infinitely small. In this way we are able to link up in a general manner mechanics with electrodynamics, to substitute energy for mass, and moreover, to reduce the building blocks of the universe from the ninety-two different atom types of the classical world picture to two – electrons and protons. Every material body consists of electrons and protons. The combination of a proton and an electron is either a neutron or a hydrogen atom, according as the electron becomes attached to the proton or circles about it. All the physical and chemical properties of a body may be deduced from the type of its structure.

The formerly accepted world picture is thus preserved, except for the fact that now it takes on the aspect of a special section of a still larger, still more comprehensive, and at the same time still more homogeneous picture. This happens in all cases within our experience. As the multitude of the natural phenomena observed in all fields unfolds in an ever richer and more variegated profusion, the scientific world picture, which is derived from them, assumes an always clearer and more definite form. The continuing changes in the world picture do not therefore signify an erratic oscillation in a zigzag line, but a progress, an improvement, a completion. In establishing this fact I have, in my opinion, indicated the basically most important accomplishment that scientific research can claim.

But what is the direction of this progress, and what is its ultimate goal? The direction, evidently, is the constant improvement of the world picture by reducing the real elements contained in it to a higher reality of a less naive character. The goal, on the other hand, is the creation of a world picture, with real elements which no longer require an improvement, and therefore represent the ultimate reality. A demonstrable attainment of this goal will – or can – never be ours. But in order to have at least a name for it, for the time being, we call the ultimate reality "the real world," in the absolute, metaphysical sense of the word, *real*. This is to be construed as expressing the fact that the *real* world – in other words, objective nature – stands behind everything explorable. In contrast to it, the scientific world picture gained by experience – the *phenomenological* world – remains always a mere approximation, a more or less well divined model. As there is a material object behind every sensation, so there is a metaphysical reality behind everything that human experience shows to be real. Many philosophers object to the word, *"behind."* They say: "Since in exact science all concepts and all measures are reducible to sensations, in the last analysis the meaning of every scientific finding also refers only to the sense world, and it is inadmissible, or at least superfluous, to postulate the existence behind this world of a metaphysical world, totally inaccessible to direct scientific inquiry and examination." The only proper reply to this argument is, simply, that in the above sentence the word, *behind*, must not be interpreted in an external or spatial sense. Instead of "behind," we could just as well say, *"in"* or *"within"* Metaphysical reality does not stand spatially *behind* what is given in experience, but lies fully *within* it. "Nature is neither core nor shell – she is everything at once," The essential point is that the world of sensation is not the only world which may conceivably exist, but that there is still another world. To be sure, this other world is not directly accessible to us, but its existence is indicated, time and again, with compelling clarity, not only by practical life, but also by the labors of science. For the great marvel of the scientific world picture, becoming progressively more complete and perfect, necessarily impels the investigator to seek its ultimate form. And since one must assume the existence of that which one seeks, the scientist's assumption of the actual existence of a "real world," in the absolute sense of the word, eventually grows into a firm conviction which nothing can shake any more. This firm belief in the absolute *Real* in nature is what constitutes for him the given, self-evident premise of his work; it fortifies repeatedly his hope of eventually groping his way still a little nearer to the essence of objective Nature, and of thereby gaining further clues to her secrets.

Since the real world, in the absolute sense of the word, is independent of individual personalities, and in fact of all human intelligence, every discovery made by any individual acquires a completely universal significance. This gives the inquirer, wrestling with his problem in quiet seclusion, the assurance that every discovery will win the unhesitating recognition of all experts throughout the entire world, and in this feeling of the importance of his work lies his happiness. It compensates him fully for many a sacrifice which he must make in his daily life.

The sublime nature of such a goal must, necessarily, dwarf into insignificance any doubt engendered by the difficulties encountered while shaping the scientific world picture. It is particularly important to emphasize this in our own day, for nowadays such difficulties are sometimes regarded as serious impediments to the salutary progiess of scientific work. It is an odd fact that experimental difficulties arc so regarded to a lesser degree than theoretical ones. The circumstance that with the increasing demands on the accuracy of measurements the instruments, too, become more intricate, is understood and accepted as a matter of course. But the fact that in the endeavor to improve continually the expansion of systematic interrelations, it is necessary to use definitions and concepts which diverge more and more from traditional forms and intuitive notions, is sometimes cited as a reproach against theoretical research, and is even viewed as indicating that theoretical research is entirely on the wrong track.

Nothing could be more shortsighted than such a view. For if we stop to think that the improvement of the world picture goes hand in hand with an approach to the metaphysically "real world," the expectation that the definitions and concepts of the objectively real world picture will not diverge too much from the framework created by the classical world picture, amounts basically to a demand that the metaphysically real world be completely intelligible in terms of ideas derived from the former naive world picture. This is a demand that can be never fulfilled. We simply cannot expect to recognize and discern the finer structure of something, so long as we flatly refuse to view it otherwise than with the naked eye. Yet, in this respect there is no reason for fear. The development of the scientific world picture is a matter of absolute necessity. The experiences gained with the refined instruments of measurement demand inexorably that certain firmly-rooted intuitive notions be abandoned and replaced by new, more abstract conceptual structures, for which the appropriate intuitions are still to be found and developed. Thus, they are the landmarks to guide theoretical research on its road from the naive concept of reality to the metaphysical "Real."

But significant as the achievements may be, and near as the desired goal may seem, there always remains a gaping chasm, unbridgeable from the point of view of exact science, between the real world of phenomenology and the real world of metaphysics. This chasm is the source of a constant tension, which can never be balanced, and which is the inexhaustible source of the insatiable thirst for knowledge within the true research scientist. But at the same time, we catch here a glimpse of the boundaries which exact science is unable to cross. May its results be ever so deep and far-reaching, it can never succeed in taking the last step which would take it into the realm of metaphysics. The fact that although we feel inevitably compelled to postulate the existence of a *real world*, in the absolute sense, we can never fully comprehend its nature, constitutes the irrational element which exact science can never shake off, and the proud name, "Exact Science," must not be permitted to cause anybody to underestimate the significance of this element of irrationality. On the other hand, the very fact that science sets its own limits on the basis of scientific knowledge itself, appears well suited to strengthen everybody's confidence in the reliability of that knowledge, knowledge obtained on the basis of incontestable presupposition and with the help of rigorous experimental and theoretical methods.

If now, we cast our glance, from the viewpoint now established, back on the starting point of our considerations, and on the entire train of thoughts pursued, the results gained will become even clearer. We began our deliberations with a definite disillusionment. We sought a universal foundation on which to erect the edifice of exact science, a foundation of indisputable firmness and security – and we failed to find it. Now in the light of the insights gained, we recognize that our quest was doomed to failure even before it started. For, basically considered, our attempt was based on the idea of starting out on our scientific exploration from something irrevocably real, whereas we have now come to understand that such ultimate reality is of a metaphysical character and can never be completely known. This is the intrinsic reason which doomed to failure every previous attempt to erect the edifice of exact science on a universal foundation valid *a priori*. We had to be satisfied, instead, with a starting point which was of inviolable solidity and yet of an extremely limited significance, since it was based solely on individual data of experience. It is at this modest point that scientific research enters with its exact methods, and it works its way step by step from the specific to the always more general. To this end, it must set and continually keep its sights on the objective reality which it seeks, and in this sense exact science can never dispense with *Reality* in the metaphysical sense of the term. But the real world of metaphysics is not the starting point, but the goal of all scientific endeavor, a beacon winking and showing the way from an inaccessibly remote distance.

The assurance that every new discovery, and every new fact of knowledge gained from it, will bring us nearer to the goal, must compensate us for the numerous, and certainly not negligible drawbacks which are necessarily created by the continual abatement of the intuitive character and ease of application of the world picture. In fact, the present scientific world picture, as against the original naive world picture, shows an odd, almost alien aspect. The immediately experienced sense impressions, the primordial sources of scientific activity, have dropped totally out of the world picture, in which sight, hearing and touch no longer play a part. A glance into a modern scientific laboratory shows that the functions of these senses have been taken over by a collection of extremely complex, intricate and specialized apparatus, contrived and constructed for handling problems which can be formulated only with the aid of abstract concepts, mathematical and geometric symbols, and which often are beyond the layman's power of understanding. One might feel completely at sea trying to puzzle out the meaning of exact science, and exact science has even been accused on this account of having lost its firm footing with the loss of its original intuitive character. But he who persists in this opinion, despite the reasons cited, is beyond help, and will be as unable to make any essential contribution to the progress of exact science as an experimenter who insists, as a matter of principle, on working always with primitive instruments only. For exact science demands more than a gift of intuition and willingness to work hard. It demands also very involved, painstaking, tedious attention to details, for which many scientists must often pool their efforts in order to prepare their branch of science for the next step on the ladder of gradual progress. To be sure, when the pioneer in science sends forth the groping feelers of his thoughts, he must have a vivid intuitive imagination, for new ideas are not generated by deduction, but by an artistically creative imagination. Nevertheless, the worth of a new idea is invariably determined, not by the degree of its intuitiveness – which, incidentally, is to a major extent a matter of experience and habit – but by the scope and accuracy of the individual laws to the discovery of which it eventually leads.

Of course, every step forward means also that the difficulty of the task increases, the demands on the analyst grow more exacting, and the need for an expedient division of labor becomes always more urgently imperative. In particular, the division of science into experimental and theoretical was completed about a century ago. Experimenters are the shock-troops of science. They perform the decisive experiments, carry out the all important work of measurement. An experiment is a question which science poses to Nature, and a measurement is the recording of Nature's answer. But before an experiment can be performed, it must be planned – the question to Nature must be formulated before being posed. Before the result of a measurement can be used, it must be interpreted – Nature's answer must be understood properly. These two tasks are those of the theorist, who finds himself always more and more dependent on the tools of abstract mathematics. Of course, this does not mean that the experimenter does not also engage in theoretical deliberations. The foremost classical example of a major achievement produced by such a division of labor is the creation of spectrum analysis by the joint efforts of Robert Bunsen, the experimenter, and Gustav Kirchhoff, the theorist. Since then, spectrum analysis has been continually developing and bearing ever richer fruit.

Whenever an experimental finding contradicts the accepted theory, another step on the ladder of progress is thereby announced, for the contradiction signifies that the accepted theory must be overhauled and improved. But the question as to just where and how to change it, entails serious difficulties. For the more tried an existing theory is, the more sensitive it is, and the stronger resistance it puts up to every attempt to alter it. In this respect, it behaves like a highly complex, widely ramified organism, whose individual component parts are mutually interdependent and are so closely interlinked that a reaction to any stimulus at any one point is also manifested automatically at quite different and, seemingly, very remote places. This gives rise to new questions, which can be investigated experimentally, and thus it may lead to consequences, the bearing and importance of which no one could suspect at the outset. This is how the theory of relativity was born, and this is the story behind the genesis of the quantum theory. In our own days, the constant growth and advancement of the youngest branch of natural science, nuclear physics, brought about and implemented by a reciprocal supplementation of experiment and theory, is another typical example of such fruitful collaboration.

IV

But why all this enormous labor, demanding the best efforts of countless soldiers of science during their entire lives? Is the ultimate result – which, as we have seen, in its individual details always drifts away from the immediately given facts of life – truly worth this costly effort?

These questions would indeed be justified if the meaning of exact science were limited to a certain gratification of man's instinctive yearning for knowledge and insight. But its significance goes considerably deeper. The roots of exact science feed in the soil of human life. But its link to it is twofold. For it not only has its source in experience, but also has a retroactive effect on human life, both material and spiritual, and the more freely it can unfold itself, the stronger and more fruitful is this effect. This manifests itself in a very peculiar manner. First, as we have found, when science works on the world picture of its own making, its quest of metaphysical Reality causes it to drift always farther and farther away from the immediate facts and interests of life, since it always takes to less intuitive and more solitary trails. But these trails, and only these, are the very paths leading toward a discernment of new laws of interrelations, which would be inaccessible in any other way, and which can then be made relevant for human experience and thus made to serve human needs.

This fact can be observed in countless individual instances. Here, too, a far-reaching division of labor has proved its worth excellently. The first step, the moulding of the world picture from its beginnings in ordinary experience, is the task of pure science. The second step, the practical utilization of the scientific world picture, is the task of technology. Both these tasks are equally important, and since either of them demands a man's full energy, if an individual scientist wants to make progress in his work, he must concentrate all his energy on one single task and for the time being forget completely other problems and interests. For this reason, never reproach the scholar too harshly for his other-worldliness and his indifference to important problems of human society. Without such a one-sided attitude, Heinrich Hertz could never have discovered radio waves, or Robert Koch the tubercle bacillus. These gifts of pure scientific research to practical life have their counterparts in the manifold stimuli and intelligent assistance which science receives from technology, a fact that is becoming progressively more manifest in our day and whose importance cannot be assessed too highly.

I feel I must discuss here a little more closely, by way of an example, a very recent and very impressive case of the often quite unsuspected close interrelation of science and technology. For a great number of years, only men of pure science were interested in the distinctive facts of atomic transformations. To be sure, the magnitude of the energies thus released did attract attention, yet since atoms were so infinitesimally small, no serious thought was given to the possibility that one day they might acquire practical significance, too. To-day, due to new findings in the field of artificial radioactivity, this question has taken an astonishing turn. The investigations of Otto Hahn and his collaborators have established the fact that a uranium atom bombarded by a neutron splits into several parts. Two or three neutrons are liberated, and each of them continues on its own path and may, in its turn, collide with a uranium atom and split it. Thus the effects may multiply; and it may happen that as a consequence of the increasing bombardment of uranium atoms by the liberated neutrons the energy thus released will swell like an avalanche within a very short time. To visualize this, think of the well known example of chain letters. With the number of available atoms, this chain reaction may reach quite enormous, hardly conceivable proportions. Of course, an indispensable prerequisite for this effect is that the free neutrons, prior to their hitting uranium nuclei, are not captured by other atoms and are thus either permanently absorbed by the latter or deflected away from uranium nuclei.

A specific computation has shown that the amount of energy released in this manner in a cubic meter (35.314 cubic feet) of powdered uranium oxide within one one-hundredth of second is sufficient to lift a weight of 1,000,000,000 metric tons to a height of almost 17 miles. This amount of energy could replace the output of all the big power plants of the world combined for many years.

Up to quite recently, a technical utilization of the energy latent in the nuclei of atoms might have appeared as a utopian dream. But it was made a reality about 1948, by the impressive collaboration of British and American scientists with American industry, backed by huge government subsidies. At the present moment, several "uranium piles" are operating in America, and the heat continually produced by them is sufficient to raise the temperature of the Victoria River in the state of Washington by 1 degree Centigrade. So far as the reports disclose, these vast amounts of energy are still unused. Right now, the problem is to get rid of them in a harmless way. But these same piles furnish also the raw materials for the atomic bombs, in which vast amounts of the nuclear energy of the atom are liberated within a fraction of a second, producing explosions beside which the devastation caused by all chemical explosives fades into insignificance. No words can be strong enough to over-emphasize the danger of self-extermination which threatens the entire human race, should a future war bring about the use of a large number of such bombs. Human imagination is incapable of conceiving the possible consequences. A particularly eloquent and forceful plea for peace is the memory of the 80,000 dead of Hiroshima and the 40,000 dead of Nagasaki, a plea addressed to all nations, and especially to their responsible leaders.

In view of these facts, perhaps many who have lost the art of wondering may feel disposed to learn it anew. And in fact, compared with immeasurably rich, ever young Nature, advanced as man may be in scientific knowledge and insight, he must forever remain the wondering child and must constantly be prepared for new surprises.

Thus we see ourselves governed all through life by a higher power, whose nature we shall never be able to define from the viewpoint of exact science. Yet, no one who thinks can ignore it. A thinking human being, who has not only scientific but also metaphysical interests, must choose one of two possible attitudes: Either fear and hostile resistance or reverence and trusting devotion. If we reflect on all the unspeakable suffering and incessant destruction of life and property which have plagued mankind since time immemorial, we may be tempted to agree with the pessimistic philosophers who consider life worthless and deny the possibility of permanent progress, of a betterment of mankind, and who profess instead that it is the fate of every human civilization to turn blindly against itself as soon as it has reached a certain peak, and to destroy itself without sense or purpose.

May exact science be cited as an evidence of such a far-reaching view? The answer must be "No," if for no other reason, because science is not qualified to decide the question. From the scientific point of view, one might just as well, and perhaps with even more justification, endorse the opposite opinion. It would require merely an extension of the range of observation, a thinking not in terms of centuries but of many millennia. Or is there anybody who would seriously deny that during the past one hundred thousand years *homo sapiens* has made progress and has improved himself? Why should this progress not continue further – if not in a straight line, then at least in waves?

Of course, such considerations, such a longrange view, are no help to the individual. They cannot bring him succor in his hour of need or cure his pain. The individual has no alternative but to fight bravely in the battle of life, and to bow in silent surrender to the will of a higher power which rules over him. For no man is born with a legal claim to happiness, success and prosperity in life. We must therefore accept every favorable decision of Providence, each single hour of happiness, as an unearned gift, one that imposes an obligation. The only thing that we may claim for our own with absolute assurance, the greatest good that no power in the world can take from us, and one that can give us more permanent happiness than anything else, is integrity of soul, which manifests itself in a conscientious performance of one's duty. And he whom good fortune has permitted to cooperate in the erection of the edifice of exact science, will find his satisfaction and inner happiness, with our great poet Goethe, in the knowledge that he has explored the explorable and quietly venerates the inexplorable.

The Concept of Causality in Physics

In the fight currently raging about the meaning and validity of the Law of Causality in modern physics, every attempt to clarify the conflicting opinions must begin with the statement that in this connection everything depends on a clear understanding of the sense in which the word "causality" is used in the science of physics. To be sure, it is agreed *a priori* that whenever a reference is made to a "causal relationship" between two successive events or occurrences, this term is understood to designate a certain regular connection between them, calling the earlier one the *cause*, and the latter one the *effect*. But the question is: What constitutes this specific type of connection? Is there any infallible sign to indicate that a happening in nature is causally determined by another?

It follows from the numerous inquiries heretofore undertaken into this question that the best and safest way to approach a clear answer is to relate the question to the possibility of making accurate predictions. In fact, there can be no more incontestable way to prove the causal relationship between any two events than to demonstrate that from the occurrence of one it is always possible to infer in advance the occurrence of the other. This point was quite familiar to the farmer who gave a visual demonstration to some sceptical peasants of the causal relationship between artificial fertilizers and the fertility of the soil, by intensively fertilizing his clover fields in certain narrow strips having the form of letters, so as to make the following sentence appear: "This strip was fertilized with calcium sulfate."

Therefore, I want to base all our subsequent considerations on the following simple proposition, equally applicable outside of the realm physics: *"An occurrence is causally determined if it can be predicted with certainty."* Of course, this sentence is meant to express only that the possibility of making an accurate prediction for the future constitutes an infallible criterion of the presence of a causal relationship, but not that it is synonymous with the latter. I need to mention merely the well known example that we can predict with a certainty while it is still day the coming of night, and yet this does not make day the cause of night.

But conversely, it also often happens that we assume the presence of a causal relationship even in cases where there is no question at all of the possibility of making accurate predictions. Just think of weather forecasts. The unreliability of weather prophets has become proverbial, and yet there is hardly any trained meteorologist who does not consider the atmospheric processes to be causally determined. All these considerations indicate that in order to find the right clues to the concept of causality, we must go still a little deeper.

In the case of weather forecasts, the thought suggests itself that their unreliability is due merely to the size and complicated nature of the object of tbe analysis, i.e. the atmosphere. If we single out a small portion of it – for instance, a cubic foot of air – we will be far more likely to make accurate predictions about its reaction to external influences, such as compression, heat, moisture, etc. We are familiar with certain laws of physics which enable us to predict, with more or less certainty, the readings of the corresponding measurements, such as the increase in pressure, rise in temperature, condensation, etc.

However, if we observe things still a little more closely,

we shall soon reach a very remarkable conclusion. Simple as we may make the conditions, and precise as our measuring instruments may be, we shall never succeed in calculating in advance the results of the actual measurements with an absolute accuracy, in other words in making the predicted value of a magnitude agree to the last decimal place with the figure actually registered by the instruments; there always remains a certain margin of uncertainty – in contrast to the calculations in pure mathematics, as in the case of the square root of 2 or of π, which can be given accurately to any number of decimal places. And whatever applies to mechanical and thermal processes, holds true for all fields of physics, including electricity and optics.

Thus, all the above cited experiences force us to recognize the following principle as a firmly established fact: *It is never possible to predict a physical occurrence with unlimited precision.* If we now compare this principle with the proposition accepted above as our starting point, namely that an occurrence is causally determined if it can be predicted with certainty, we find ourselves facing an unavoidable dilemma: We may elect *either* to adhere literally to the exact wording of our basic proposition, in which case there cannot exist even one single instance in nature where a causal relationship would have to be assumed to prevail – *or* to subject that basic proposition to a certain modification, so designed as to provide room for the presupposition of strict causality.

A number of contemporary physicists and philosophers have chosen the first alternative. I shall refer to them here as *indeterminists.* They claim that genuine causality, strict regularity, actually does not exist in nature but is merely an illusion created by the operation of certain rules which are never of an exact universal validity, even though they often come very near to it. Upon closer consideration, the indeterminist discovers a statistical root in every law of physics, including the law of gravity and of electrical attraction; he regards them, one and all, as laws of probability, relating only to mean values of many similar observations, possessing only an approximate validity for individual instances.

A good example of such a statistical law is the dependence of the magnitude of gas pressure on the density and temperature. The pressure of a gas is produced by the continual impact against the walls of the vessel of a vast number of gas molecules moving at random in all directions with great velocity. A summary computation of the aggregate dynamic effect of their impact reveals that the pressure against the walls of the vessel is almost proportional to the density of the gas as well as to the mean square of the molecular velocity, a result which is in satisfactory agreement with measurement if we regard the temperature as a measure of molecular velocity.

A direct confirmation of this theory of gas pressure is furnished by investigations on the temporal variations of the pressure against a very small portion of the wall of the vessel. Such variations, produced by random molecular impacts, can be observed wherever molecules in rapid flight come in contact with easily movable bodies; they manifest themselves also in the Brownian molecular movement, as well as in the fact that very sensitive scales never come to a complete rest, but constantly execute minute irregular vibrations about their position of equilibrium.

Analogously with the gas laws, the indeterminists attribute every other kind of physical regularity, ultimately to the operation of chance. They see nature ruled exclusively by statistics, and their aim is to base all physics on the calculus of probabilities.

Actually, physical science has developed up to now on the very opposite foundation. It chose the *second* one of the two alternatives mentioned above: In order to be able to maintain the full and absolute validity of the law of causality, it modified slightly the basic proposition, that an occurrence is causally determined if it can be predicted with a certainty.

This was done by using the word "occurrence" in a somewhat modified sense. Thus, theoretical physics considers as an occurrence not an actual individual process of measurement – which always includes accidental and unessential elements, too – but a certain, merely theoretical process; and in this manner it replaces the sense world, as given to us directly by our sense organs (or alternatively, by the measuring instruments which serve us as sharpened sense organs), by another world, the world picture of physics, which is a conceptual structure, arbitrary to a certain degree, created for the purpose of getting away from the uncertainty involved in every individual measurement and for making possible a precise interrelation of concepts.

Consequently, every measurable physical magnitude, every length, every time interval, every mass, every charge, has a double meaning, according as we regard it as directly given by some measurement, or conceive of it as translated into the world picture of physics. In the first interpretation, it is never capable of a sharp definition, and can therefore never be represented by a quite definite number; but in the world picture of physics it stands for a certain mathematical symbol, which lends itself to manipulation according to quite definite, precise rules. This goes for the height of a tower just as for the duration of the swing of a pendulum or for the brightness of an incandescent lamp. A clear and consistent distinction between the magnitudes of the sense world and the corresponding magnitudes of the world picture is absolutely essential for the clarification of concepts; without such a distinction it is impossible to discuss these questions intelligently and objectively.

It is absolutely untrue, although it is often asserted, that the world picture of physics contains, or may contain, directly observable magnitudes only. On the contrary, directly observable magnitudes are not found at all in the world picture. It contains symbols only. In fact, the world picture even contains constituents which have only a very indirect significance for the sense world, or no significance at all, such as ether waves, partial vibrations, frames of reference, etc. Primarily, such constituents play the part of dead weight or ballast, but they are incorporated because of the decisive advantage assured by the introduction of the world picture – that it permits us to carry through a strict determinism.

Of course, the world picture always remains a mere auxiliary concept. It is self-evident that in the last analysis, the things that really matter, are the occurrences in the sense world and the greatest possible accuracy in predicting them. In classical physics this is achieved as follows: First, an object found in the sense world – for instance, a system of material points – is symbolically represented in some measured condition, i.e. is translated into the world picture. One thus obtains a certain physical structure in a certain initial state. Similarly, suitable symbols are substituted in the framework of the world picture for external influences which operate subsequently on the object. One thus obtains the external forces acting on the structure, and the corresponding boundary conditions. The behavior of the structure is then unambiguously determined for all times by these data, and it can be computed with absolute accuracy from the differential equations of the theory. The coordinates and momenta of all material points of the structure are thus exhibited as quite definite functions of time. If then at any later time the symbols used in the world picture are retranslated into the sense world, one thus obtains a connection between a later occurrence in the sense world and a previous occurrence in the sense world, and this connection can be utilized for an approximate prediction of the later occurrence.

In summary, we may say: While in the sense world the prediction of an occurrence is always associated with a certain element of uncertainty, in the world picture of physics all occurrences follow one another in accordance with precisely definable laws – they are strictly determined causally. Therefore, the introduction of the world picture of physics – and herein lies its significance – reduces the uncertainty of predicting an occurrence in the sense world to the uncertainty in translating that occurrence from the sense world into the world picture and in retranslating it from the world picture into the sense world.

Classical physics was but little concerned with this uncertainty; its main concern was to follow through the causal point of view in the consideration of the occurrences in the world picture, and this was where it achieved its great results. Specifically, it succeeded also in finding a satisfactory interpretation, on the basis of a strict causality, for the above mentioned irregular vibratory phenomena corresponding to the Brownian molecular movement. The indeterminists see no real problem here. For since they look for irregularity behind every rule, statistical regularity is that which gives them direct satisfaction. Therefore, they are satisfied also with the assumption that the collision of two individual molecules, as well as the impact of the molecules against the wall of the vessel, occurs solely according to statistical laws. Nevertheless, this assumption is as little justified as would be the conclusion that the charge of an individual electron is located on its surface just because in a charged conductor the electrons are all located on its surface. On the other hand, the determinists who, on the contrary, look for a rule behind every irregularity, were led to the problem of basing a theory of the gas laws on the premise that the collision of two individual molecules is determined in a strict causal manner. The solution of this problem was the life work of Ludwig Boltzmann, and it represents one of the most beautiful triumphs of theoretic research. For it not only yields the principle, confirmed by actual measurements, that the mean energy of the oscillations about the state of equilibrium is proportional to the absolute temperature, but it permits also a remarkably accurate computation of the absolute number and mass of the molecules, based on a measurement of these oscillations, as in the case of a highly sensitive torsion balance.

Such outstanding achievements seemed to justify the hope that the world picture of classical physics would in principle accomplish its task, and that a steady improvement and refinement of the technique and methods of measurement would progressively reduce the significance of the uncertainties accompanying the translation into and from the sense world. But the introduction of the elementary quantum of action destroyed this hope at one blow and for good.

The so-called Uncertainty Principle, discovered and formulated by Heisenberg, constitutes a characteristic feature of quantum mechanics; it asserts that for any two canonically conjugate magnitudes, such as position and momentum, or time and energy, only one can be measured to any desired degree of accuracy, so that an increase in the precision of the measurement of one magnitude is accompanied by a proportional decrease in the precision of measurement of the other. Consequently, when one magnitude is ascertained with absolute accuracy, the other one remains absolutely indefinite.

It is evident that this principle fundamentally precludes the possibility of translating into the sense world, with an arbitrary degree of accuracy, the simultaneous values of the coordinates and momenta of material points, as these are conceived in the world picture of classical physics; this circumstance constitutes a difficulty with respect to the recognition of a universal validity of the principle of strict causality, and it has even caused some indeterminists to regard the law of causality in physics as decisively refuted. However, upon closer scrutiny, this conclusion – founded on a confusion of the world picture with the sense world – proves a rash one, to say the very least. For there is another, more logical way out of the difficulty, a way which has often rendered excellent services on previous occasions – namely, the assumption that the attempt to determine simultaneously both the coordinates and the momentum of a material point is physically completely meaningless. However, the impossibility of giving an answer to a meaningless question must, of course, not be charged up against the law of causality, but solely against the premises which produced that particular question; in other words, in this particular instance, against the assumed structure of the world picture of physics. And since the classical world picture has failed, another must take its place.

This is what actually has happened. The new world picture of quantum mechanics is a product of the need to find a way of reconciling the quantum of action with the principle of strict determinism. For this purpose, the traditional primary constituent of the world picture, the material point, had to be deprived of its basic, elementary character; it was resolved into a system of material waves. These material waves constitute the primary elements of the new world picture. The material point in its old meaning now appears merely as a special borderline case, as an infinitesimally small parcel of waves, the momentum of which is totally indefinite, since its position is definite, according to Heisenberg's uncertainty principle. If we assign a certain range to the position of the material point, the value of the momentum also becomes approximately definite and the laws of classical mechanics then are approximately valid for positions and momenta.

In general, the laws of material waves are basically different from those of the classical mechanics of material points. However, the point of central importance is that the function characterising material waves, i.e. the wave function or probability function (the term itself is of no importance here), is fully determined for all places and times by the initial and boundary conditions, according to quite definite principles of computation, whether we use Schroedinger's operators, Heisenberg's matrices, or Dirac's q numbers.

We thus see that the principle of determinism is as strictly valid in the world picture of quantum mechanics as in that of classical physics. The difference consists only in the symbols used and in the mathematics applied. Accordingly, in the realm of quantum mechanics, just as formerly in classical physics, the uncertainty in the prediction of the occurrences of the sense world is reduced to an uncertainty in the correlation of world picture and sense world, in other words, to an uncertainty in the translation of the symbols of the world picture into the sense world, and *vice versa*. The fact that this double uncertainty is involved is the most impressive proof of the importance of retaining the principle of determinism within the world picture.

Nevertheless, even a cursory glance shows how far the world picture of quantum mechanics has shifted from the sense world, and how much more difficult it is to translate an occurrence from the world picture of quantum mechanics into the sense world or *vice versa*, than was the case in classical physics. In the latter, the meaning of every symbol was immediately and directly intelligible; the position, the velocity, the momentum, the energy of a material point could be determined more or less accurately by measurements, and there appeared no obvious reason to doubt that a steady improvement of the technique and methods of measurement could not eventually reduce the remaining factor of uncertainty below any desired margin. On the other hand, the wave function of quantum mechanics supplies no direct clue whatever for any obvious interpretation of the sense world, simply because it does not refer to ordinary space at all but rather to the configurational space which has as many dimensions as there are coordinates present in the physical structure under consideration. Moreover – and this is where the real trouble lies – the wave function does not give us the values of the coordinates as functions of the time, but merely the probability that the coordinates may possess certain specific values at some specified time.

The indeterminists again seized on this circumstance as the occasion for a new attack against the law of causality. And this time their attack actually seems to promise them success; for on the basis of measurements it is possible to assign merely a statistical significance to the wave function. Nonetheless, once more the defenders of strict causality have the same way out as before: The assumption is that the question concerning the meaning of a certain symbol of the world picture of quantum mechanics, such as a material wave, has no definite sense unless one also specifies the condition of the particular measuring apparatus used for translating the symbol into the sense world. For this reason, one is led to talk about the causal effect of the measuring instrument that is employed; and one is thereby expressing the fact that the particular uncertainty in question is determined at least in part by the circumstance that the magnitude of the value to be measured is dependent in a certain regular manner on the method used for measuring it.

However, this auxiliary hypothesis shunts the entire question to a track, the further course of which is still hidden in darkness. For now the indeterminists are justified in raising the question whether the concept of a causal influence of the measuring instrument on the measuring process has any intelligible meaning, since every attempt to examine such a causal effect directly requires some kind of measurement, and since with every new measurement a new causal influence and therefore a new uncertainty would be introduced into the problem.

And yet, this objection still does not dispose of the matter. For as every experimental physicist knows, besides the direct methods of investigation there exist also indirect ones, and the latter accomplish good results in many a case where the former have failed. However, I must take exception to the view (a very popular one these days, and certainly a very plausible one on the face of it) that a problem in physics merits examination only if it is established in advance that a definite answer to it can be obtained. If physicists had always been guided by this principle, the famous experiment of Michelson and Morley undertaken to measure the so-called absolute velocity of the earth, would never have taken place, and the theory of relativity might still be nonexistent. Accordingly, if the study of a question now regarded fairly universally as meaningless, such as that of the absolute velocity of the earth, has turned out to produce such extraordinary benefits to science, how much more worth while must it be to follow up a problem, the deeper meaning of which is still under debate and which is capable more than any other of enriching research.

But how are we to come to a decision? Obviously, there is no other alternative than to consider the two opposing views, to side with the one that appeals to us more, and then to investigate whether it leads to valuable or to worthless conclusions. At any rate, one must welcome the fact that physicists who are closely interested in this subject are split into two factions; one leans toward the theory of determinism, the other toward indeterminism. So far as I can see, the latter represent the majority nowadays, although it is difficult to establish the facts which may easily change in the course of time. Between these two viewpoints there is also room for a third position which in a certain sense is a mediating one, in that it assigns to certain concepts, such as the force of electrical attraction, a direct significance and a strict uniformity with respect to the sense world; whereas to other concepts, such as material waves, it assigns merely a statistical significance. However, because it lacks systematic unity this view does not appear to be very satisfactory. For this reason, I propose to disregard it now and to confine myself to a somewhat closer analysis of the two fully consistent viewpoints.

The indeterminist is satisfied in his quest for knowledge by the discovery that the wave function of quantum mechanics is merely a probability value. He has no further problem in connection with it. On the other hand, he sees unsolved problems in certain determinate laws of nature, such as Coulomb's law of electrical attraction; for he cannot be satisfied with Coulomb's formula for the force or the potential, but continues to search for exceptions, and is not content until he succeeds in determining the magnitude of the probability that the electrical force differs from Coulomb's value by some arbitrary pre-assigned value.

The determinist thinks along the very opposite lines in all these points. He assigns to Coulomb's law the satisfactory character of absolute validity. On the other hand, he recognizes the wave function as a mere probability magnitude only so long as the measuring instrument used in the study of the wave is disregarded, and he looks for inflexible laws connecting the properties of the wave function and the occurrences in the measuring apparatus. For this purpose, he must obviously include in the subject matter of his investigation both the wave function and the measuring instruments. In other words, he must translate into his physical world picture not only the entire experimental setup used for the production of the material waves – such as the high voltage battery, the incandescent filament, the radioactive compound – but also the measuring apparatus, such as the photographic plate, the ionization chamber, the point counter, together with all the processes occurring therein, and he must deal with all these objects as one single structure, as an isolated system. But this does not yet take care of the problem. On the contrary – it makes it even more complicated. For in this case, if the total structure is to retain its unique character, it must neither be split up nor exposed to external influences, so that a direct examination is completely impossible. Nevertheless it now becomes possible to formulate certain new hypotheses concerning the internal processes of the system, and subsequently to test their consignments. Whether this procedure leads to any actual advance, is a question which only the future can answer. For the time being, it is still impossible to determine clearly what direction will lead to progress. At any rate, all the circumstances mentioned prove that the elementary quantum of action erects an objective barrier which limits the efficiency of the measuring apparatus available to physical science, and that, therefore, the desired progress will only give this barrier even sharper outlines than it had before.

Properly speaking, we have thus reached the end of our considerations which have demonstrated that an adherence to a strictly causal outlook – always taking the word "causal" in the modified sense explained previously – is by no means excluded from the viewpoint of even modern physics, though its necessity can never be proved either *a priori* or *a posteriori*. Nevertheless, not even a convinced determinist – indeed, especially not a convinced determinist – can escape at this point some doubts which prevent him from accepting as fully satisfactory, the interpretation of causality indicated above. For even if it should prove to be feasible to develop further the concept of causality along the lines described, the concept as here proposed involved a serious flaw. For it could be maintained that a relation possessing such profound significance as the causal connection between two successive events represents ought to be independent by its very nature from the human intellect which is considering it. Instead, we have not only linked at the very outset the concept of causality to the human intellect, specifically to the ability of man to predict an occurrence; but we have been able to carry through the deterministic viewpoint only with the expedient of replacing the directly given sense world by the world picture of physics, that is, by a provisional and alterable creation of the human power of imagination. These are anthropomorphic traits which ill-befit fundamental concepts of physics; and the question therefore arises whether it is not possible to give the concept of causality a deeper meaning by divesting it as far as it can be of its anthropomorphic character, and to make it independent of human artifacts such as the world picture of physics. Of course, we shall have to adhere to our basic premise, that an occurrence is causally determined if it can be predicted with a certainty, for otherwise we would lose the only solid foundation for our discussion. And we must feel bound to no less a degree to our second principle, that it is never possible, to predict an occurrence with absolute precision. It accordingly follows that if we are to speak at all of causality in nature, we must introduce some modification into our first basic proposition. To this extent the situation remains as it was before. But the type of modification which we introduced previously into the first basic proposition can be replaced by a totally different one.

In the preceding discussion what was modified was the object of the prediction – the occurrence. For we related the occurrences not to the directly given sense world, but to an artificially created world picture, and we thus found it possible to determine occurrences accurately. But instead of modifying the object of predictions we can alter our notion of the subject of the prediction, that is, of the predicting intellect. For every prediction presupposes somebody who does the predicting. Therefore, in the following discussion we shall direct our attention solely to the predicting subject, and shall consider as the objects of prediction the directly given occurrences of the sense world, without introducing an artificial world picture.

First of all, it is easy to see that the reliability of a prediction depends, to a high degree, on the individual personality of the one who is making it. If we consider again weather forecasts, it makes a great difference whether tomorrow's weather is forecast by an amateur who knows nothing about today's atmospheric pressure, wind direction, atmospheric temperature and humidity, .or by a capable farmer who considers all these data and also has a great deal of experience, or finally, by a scientifically trained metereologist who in addition to the local data also has access to the accurate information supplied by a great many weather-maps from near and far. The forecast of the experienced farmer is more reliable than that of the amateur, and the forecast of the trained metereologist more reliable than that of either of those two. In view of this circumstance, it seems natural to suppose that an ideal intellect, intimately familiar with the most minute details ol physical processes occurring concurrently everywhere, would be able to predict tomorrow's weather in all its details with an absolute accuracy. The same idea applies to every other prediction of physical occurrences.

Such an assumption signifies an extrapolation which cannot be demonstrated logically, though it cannot be refuted in an *a priori* fashion either; it must therefore be judged not on the score of its truth, but only on the basis of its value. In consequence the actual impossibility of predicting ever a single occurrence accurately in classical as well as in quantum physics, appears to be a natural consequence of the circumstance that man with his sense organs and measuring instruments is himself a part of nature, subject to its laws and confined within its limits, whereas the ideal intellect is free of all such limitations.

However, in order to be able to follow through such a view logically, we must comply with an important requirement: We must be on our guard against the temptation to make the ideal intellect the object of a scientific analysis, to regard it as something analogous to ourselves, and to ask of it how it obtains the knowledge which enables it to make precise predictions. The inquisitive human being who would do so, is quite likely to hear this answer to his question: "You resemble the intellect which you comprehend, not *me*." And if after this reprimand he persists in his obstinacy and declares the concept of an ideal intellect to be meaningless and unnecessary, if not illogical, let him be reminded that not all statements which lack a logical foundation are scientifically worthless, and that his short-sighted formalism stops up the very fountain at which a Galileo, a Kepler, a Newton, and many other great physicists have quenched their thirst for scientific knowledge and insight. For all these men devotion to science was, consciously or unconsciously, a matter of faith – a matter of a serene faith in a rational world order.

Of course, this faith can no more be forced upon anybody than could one be commanded to see the truth or forbidden to commit an error. But the plain fact that we are able, at least to a certain degree, to subject future natural occurrences to our thought processes and to bend them to our will, would be a totally incomprehensible mystery, did it not permit us to surmise at least a certain harmony between the external world and the human intellect. The depth to which one conceives this harmony to extend is a matter of merely secondary importance.

In conclusion we may therefore say: The law of causality is neither true nor false. It is rather a heuristic principle, a signpost – and in my opinion, our most valuable signpost – to help us find our bearings in a bewildering maze of occurrences, and to show us the direction in which scientific research must advance in order to achieve fertile results. The law of causality, which immediately impresses the awakening soul of the child and plants the untiring question *"Why?"* into his mouth, remains a lifelong companion of the scientist and confronts him incessantly with new problems. For science is not a contemplative repose amidst knowledge already gained, but is indefatigable work and an ever progressive development.

Religion and Natural Science

A Lecture, delivered in May 1937.

In former days, when a natural scientist had to address a general audience of laymen on a subject taken from his own special field of activity, in order to awaken a certain interest in the minds of his listeners, he would be forced to link his discourse to certain palpable experiences and views of daily life, in the fields of technology, metereology or biology, and to use these as his starting points to explain the methods applied by science in order to push forward from concrete individual problems to a knowledge of universal laws. Not so today. The exact methodology now employed by natural science has proved to be so extraordinarily productive in the course of centuries that natural-scientific research nowadays dares approach also problems intuitively less obvious than those lying within the fields just mentioned, and is able to tackle successfully also problems in psychology, in epistemology, indeed even in general attitudes toward life, thereby subjecting these problems to a treatment that is thorough from its own point of view. We may justly say that in these days no question, be it ever so abstract, can arise in our civilization without being related, in one way or another, to a problem that can be handled by the methods of natural science.

Accordingly, I will not appear to be too bold as a student of nature in discussing religious problems. This is a subject, the significance of which for our entire civilization is becoming

125

progressively more manifest and which will undoubtedly be of a decisive importance for the question as to the fate that awaits us.

I

"Tell me – how do you stand to religion?" – If Goethe's Faust contains at all a simple phrase that captivates even a sophisticated listener and arouses a hidden tension within him, it must be this worried question of an innocent girl, in fear for her newly-found happiness, to her lover whom she recognizes as a higher authority. For this very same question is the one which from time immemorial has innerly moved and worried countless human beings in search of peace of mind and knowledge at the same time.

But Faust, slightly embarrassed by this candid question, can think offhand only of this mildly defensive reply: *"I want to deprive nobody of his sentiments and his church."*

I could choose no better phrase to introduce our subject. I have not the slightest intention to loosen the foundation under the feet of those among you who have made peace with their conscience and have gained that firm foothold which is a prime requisite of one's conduct in life. To do so would be a reprehensible undertaking, unfair both to those who feel so secure in their religious faith that they do not permit a natural scientific knowledge to influence it in any manner whatever, as well as to those who feel no need for any particular religious activity and are fully contented with an intuitive ethics. But these latter are likely to represent a small minority only. For the history of all eras and races teaches us only all too impressively that the candid faith which nothing can confuse, such as that which religion instills in its followers who are busy in active life, is the very fountainhead of the mightiest incentives to significant creative achievements, in the field of politics no less than in the realms of art and science.

This candid faith – and let us not delude ourselves about this – no longer exists today, even among the great masses of the nation, nor can it be revived any longer by considerations and measures oriented toward the past. For "to believe" means "to recognize as a truth," and the knowledge of nature, continually advancing on incontestably safe tracks, has made it utterly impossible for a person possessing some training in natural science to recognize as founded on truth the many reports of extraordinary occurrences contradicting the laws of nature, of miracles which are still commonly regarded as essential supports and confirmations of religious doctrines, and which formerly used to be accepted as facts pure and simple, without doubt or criticism.

Therefore, he who is in earnest about his faith and cannot bear to see it conflict with his scientific learning, must decide in his conscience whether in all honesty, he may still consider himself a member of a religious community whose creed incorporates a belief in miracles.

For a while many could still find a certain temporary reassurance by trying to steer a middle course and limiting themselves to accepting as true a few miracles of especial importance. But such an attitude is untenable in the long run. The faith in miracles must yield ground, step by step, before the steady and firm advance of the forces of science, and its total defeat is indubitably a mere matter of time. The young generation of our own era, which in any case is sharply critical, toward traditional views, no longer permits itself to be bound innerly by doctrines which it regards as contradictory to the laws of nature. And the spiritually most gifted members of the young generation in particular, those destined to be the future leaders of their nation and who not seldom harbour a burning desire for religious satisfaction, are the ones most painfully hit by such incongruities. They are the ones who must suffer most heavily if they are sincere in seeking a compromise between their religious and their scientific beliefs.

Under these circumstances, it is no wonder that the atheist movement which calls religion an arbitrary delusion invented by power-hungry priests and which has nothing but words of derision for the pious faith in a supreme power above man, is eagerly taking advantage of the progress of scientific knowledge; allegedly in alliance with natural science, the movement continues to spread at an ever quickening pace its disruptive influence over all nations and classes of mankind. I need not go here into a more detailed discussion of the fact that the victory of atheism would not only destroy the most valuable treasures of our civilization, but – what is even worse – would annihilate the very hope for a better future.

Thus, Marguerite's question to the man to whom she gave her love and trust gains a most profound significance also for those who anxiously endeavor to find out whether the progress of natural sciences is actually bringing about a destruction of true religion.

If we study Faust's concise reply, spoken with all care and tenderness of feeling, we find that we cannot give it here as our own, for a double reason: First, we must remember that this reply, both in form and content, is designed for the power of comprehension of a simple uneducated girl, and is therefore not meant to impress the intellect as well as the emotions and the imagination. But then – and this consideration is of a more decisive importance – we must bear in mind that these words are spoken by a Faust ruled by sensual desire, a confederate of Mephistopheles. I am sure that the redeemed Faust, whom we meet at the end of the second part, would give a somewhat different answer to Marguerite's question. But I do not presume to conjecture on the secrets which the poet chose to keep permanently as his own. I prefer to attempt to cast some light, from the perspective of one trained in the spirit of exact scientific research, on the question of whether and to what extent a truly religious attitude is compatible with the facts of knowledge gained through natural science – or to express it more concisely: Whether a person trained in natural science can be truly religious at the same time. For this purpose, first of all let us discuss two special questions, quite separately from each other. The first one is: What demands does religion make on the beliefs of its followers, and what are the characteristics of a genuine religious attitude? The second question is: What is the nature of the laws natural science teaches us, and what truths does it regard as indubitable?

Once we shall have answered these two questions, we shall be in the position to decide whether and to what extent the demands of religion are compatible with those of natural science, and therefore, whether religion and natural science can exist side by side without clashing with each other.

II

Religion is the link that binds man to his God. It is founded on a respectful humility before a super-natural power, to which all human life is subject, and which controls our weal and woe. To be in harmony with this power and to enjoy its good graces, is the incessant endeavor and supreme goal of the religious person. Only in this way can he feel protected from the foreseen and unforeseen dangers, which threaten him in this earthly life, and can he enjoy that purest of all happiness, the inner peace of mind and soul that is secured only by a firm link to God and by an unconditionally trusting faith in His omnipotence and benevolence. In this sense, religion is rooted in the consciousness of the individual.

But its significance transcends the individual. Instead of each individual possessing his own distinctive religion, religion seeks to become valid and meaningful for a larger community, for a nation, for a race, and ultimately for all of mankind. For God is the sovereign of every country on this earth; the whole world with all its treasures and all its horrors is subject to Him, and there is no portion either of the realm of nature or of the mind without His omnipresence.

Therefore, the spirit of religion unites its adherents in a universal alliance, and sets before them the task of mutually acquainting each other with their articles of faith and giving them a common manifestation. But this can be accomplished only by clothing the substance of religion in a definite external form, suitable because of its intuitive clarity for the creation of a mutual understanding. In view of the great diversity of the races of man and of their ways of life, it is only natural that this external form is quite different in different parts of the world, so that a large variety of religions have come into existence in the course of the ages, A common feature of all of them consists in the rather natural assumption of a personified or at least an anthropomorphic deity. This leaves room for the most diverse concepts of the attributes of God. Each religion has its own distinct mythology and also its own distinct rituals, elaborate to the most minute details in the more highly developed religions. These are the source of certain interpretive symbols of religious worship, which are capable of acting directly on the imagination of the great masses, arousing their interest in religious matters and giving them a certain understanding of the deity. Thus, a systematic unification of mythological traditions and a strict observance of solemn ritualistic customs invest the worship of God with an external symbolical form, and centuries of incessant observance and systematic education of generation after generation increase the significance of such religious symbols. The holiness of an unfathomable deity is translated into the holiness of intelligible symbols. They are a source of strong stimulation for the arts as well. In fact, the mightiest benefits ever enjoyed by art were the result of its becoming a servant to religion.

Yet, a careful distinction must be made here between art and religion. A work of art carries its significance essentially within itself. Even though, as a rule, it owes its origin to external circumstances and in consequence often awakens trains of thought moving at a tangent, still it is basically self-sufficient and requires no specific interpretation in order to be appreciated. This fact becomes most clearly evident in music, most abstract of all arts.

On the other hand, a religious symbol always points beyond itself. Its significance is never exhausted by its own features, however much veneration it may enjoy because of its own age and the operation of a pious tradition. It is important to emphasize this because the development of civilization makes the high esteem enjoyed by certain religious symbols subject to certain inevitable changes in the course of the centuries, and it is in the interest of a genuine spirit of religion to establish the fact that what stands behind and above these symbols is unaffected by such changes.

To cite here just one of a great many specific examples: A winged angel has been regarded from time immemorial as the most beautiful symbol for a servant and messenger of God. But there can be found among persons trained in anatomy, some whose scientifically conditioned imagination does not permit them, despite their best intentions, to see any beauty in such a physiological impossibility. Nevertheless, this circumstance need have not the slightest adverse effect on their religious convictions. They ought, however, to be on their guard not to impair or destroy the pious attitudes of those who still find solace and edification in the sight of a winged angel.

But the overrating of the significance of religious symbols opens the gates to another, far more serious danger of an onslaught by the atheistic movement. It is one of the favorite techniques of the atheists, whose aim is the undermining of every true religious feeling, to direct their attacks against old-established religious rites and customs and to hold them up to ridicule or contempt as outmoded anachronisms. Through such attacks against symbols, they expect to hurt religion itself, and the more strange and conspicuous those views and customs are, the easier it is for the atheists to score a success. Many a religious soul has succumbed to these tactics.

Against this peril there is no better defense than to understand clearly and thoroughly that a religious symbol, be it ever so venerable, never represents an absolute value but is always only a more or less imperfect sign of something higher and not directly accessible to human senses.

Under these circumstances, it is quite understandable that the history of religions records the frequent recurrence of the idea to restrict or even eliminate completely the use of religious symbols and to treat religion more as a matter of abstract reasoning. But even a brief reflection shows such an idea to be entirely inadequate. Without symbols, human beings could not communicate with each other at all. This applies not only to religious communication, but to all human transactions in secular daily life as well. Language itself is actually nothing else than a symbol for something higher – for thought. To be sure, individual words also arouse a typical interest in themselves, but viewed more closely, a word is just a series of letters; its meaning lies fundamentally in the concept which it expresses. And it is basically of a secondary importance whether the concept is represented by one word rather than by another, in one particular language rather than another. If the word is translated into another language, the concept itself remains unaffected.

Another example: A flag is the symbol of the glory and honor of a regiment of soldiers. The older the flag, the higher its value. In the heat of the battle, the bearer of the flag considers it to be his supreme duty not to desert the honored emblem at any cost, to protect it with his own body if need be, even to give his life for it if he must. Yet, a flag is just a symbol, a piece of a bright-colored cloth. The enemy can capture it, soil, mutilate it, but can never destroy the higher concept, of which it is the symbol. The regiment will still retain its honor, get a new flag, and perhaps exact suitable retribution for the insult to its emblem.

Just as in an army and more generally in every community of men facing great tasks together – religion also finds symbols and their corresponding ecclesiastical rituals absolutely indispensable. They signify the highest and most venerated of all the products of human imagination directed heavenward. But we must never forget that even the most sacred symbol is of human origin.

Had mankind taken this truth to heart at all times, it would have been spared an infinity of woe and suffering. For the terrible religious wars, the horrible persecutions of heretics and their attendant tragic consequences, are in the last analysis the outcome of conflicts between opposing propositions, each possessing a certain validity and each originating in the circumstances that a common abstract idea, such as the belief in an omnipotent God, was confused with its visible but distinct media of expression, such as ecclesiastic articles of faith. Certainly there is nothing more distressing than the bitter fight of two adversaries, each of whom is fully convinced of the rightness of his cause and is filled with honest enthusiasm for it, feeling that he must devote all his energy to the battle, even sacrifice his life in it. How much productive work could have been accomplished if in the domain of religious activity such valuable energies had been united instead of employed for mutual extermination.

The deeply religious individual who gives expression to his belief in God through a veneration of his beloved sacred symbols, does not insist on them blindly, but has an understanding for the fact that there can be other persons as deeply religious as himself to whom other symbols are no less beloved and sacred – just as a definite concept remains unaltered whether it is expressed by one word or another, in one language or another.

But a comprehension of this state of affairs still does not explain the, nature of the characteristic features of true religious conviction. For now another question, the one of the truly fundamental significance, must be answered: Does that higher power which stands behind the religious symbols and lends them their essential significance, dwell solely in the human mind, and is it obliterated also in the moment when that mind ceases to exist, or does it stand for something more? In other words: Does God live in the soul of the believer only, or does He rule the world independently of whether or not one believes in Him? This is the point at which minds part company basically and decisively. This question can never be cleared up scientifically, by logical conclusions based on facts. The answer is solely and exclusively a matter of faith – religious faith.

The answer of the religious person is that God exists, that He existed before there were human beings on the earth, that he holds the whole world, believers as well as disbelievers, in His omnipotent hands since the beginning of eternity, and that He will continue to rule from His heights inaccessible to human imagination long after the earth and everything on it will have crumbled to dust. All those who profess this belief and, filled by it with humility and unquestioning trust and devotion, feel protected by the Almighty from every danger in life, all those – but only those – may consider themselves truly religious.

This is the essential content of creed which religion requires its followers to profess. Let us now see whether and how these requirements are compatible with those of science, natural science in particular.

III

In proceeding to examine what laws science teaches us, and what truths it considers to be inviolable, our task will be simplified and our purpose fully served by keeping to the most exact of all natural sciences, physics. For this is the branch of science which by every logic could be expected to be the most likely to conflict with the demands of religion. Therefore, we are to inquire what kind of discoveries physical science has made up to our most recent days, and what limits these might set for religious faith.

I hardly need to point out that viewed historically and on the whole, the findings of physical research and the conclusions resulting from them do not exhibit a purposeless change, but have been steadily improving in precision and completeness until the most recent days, some times at a faster at other times at a slower rate; and we therefore have every reason to regard the knowledge thus far accumulated by physical science as being of a lasting character.

What then is the substance of these findings? First of all, it must be noted that all data of physical knowledge are founded on measurements, and that every measurement takes place in space and time with the orders of magnitudes varying from the inconceivably vast to the infinitesimally small. We can get an approximate idea of the distance of the cosmic regions from which a message can still reach us if we consider that, light, which traverses the distance from the Moon to the Earth in a second or so, requires many millions of years to arrive from those regions to our planet. On the other side of the picture, physical science must calculate with spatial and temporal magnitudes, the infinite smallness of which can be realized by the comparison of a head of a pin with our entire planet.

Measurements of the most varied kinds have been consistent in leading to the conclusions that all physical occurrences without exception can be reduced to mechanical or electrical processes, produced by the movements of certain elementary particles, such as electrons, positrons, protons and neutrons; and both the mass and the charge of all of these particles are each expressed by an extremely small but quite definite number. The precision of this number increases with improvements in the accuracy of methods of measurement. These minute numbers, the so-called universal constants, are in a sense the immutable building blocks of the edifice of theoretic physics.

So now we must continue with the question: What is the real meaning, of these constants? Are they, in the last analysis, inventions of the inquiring mind of man, or do they possess a real meaning independent of human intelligence?

The first of these two views is professed by the followers of Positivism, or at least by its most extreme partisans. Their theory is that physical science has no other foundation than the measurements on which its structure is erected, and that a proposition in physics makes any sense only insofar as it can be supported by measurements. But since every measurement presupposes an observer, from the positivistic viewpoint the real substance of a law of physics can never be detached from the observer, and it loses its meaning as soon as one attempts mentally to eliminate the observer and to see something more, something real, behind him and his measurement.

This outlook cannot be challenged from the purely logical viewpoint. And yet, a closer examination must brand this version of it as inadequate and unproductive. For it disregards a circumstance which is of a decisive importance in the extension and progress of scientific knowledge. However much Positivism may regard itself as proceeding without presuppositions, it is nonetheless committed to a fundamental premise if it is not to degenerate into an unintelligible solipsism. This premise is that every physical measurement can be reproduced, so that its outcome depends neither on the personality of the individual performing the experiment, nor on the place and time of the measurement, nor on any other attendant circumstance. But this simply means that the factor which is decisive for the result of the measurement lies beyond the observer, and that one is therefore necessarily led to questions concerning real causal connections operating independently of the observer.

To be sure, it must be agreed that the positivistic outlook possesses a distinctive value; for it is instrumental to a conceptual clarification of the significance of physical laws, to a separation of that which is empirically proven from that which is not, to an elimination of emotional prejudices nurtured solely by customary views, and it thus helps to clear the road for the onward drive of research. But Positivism lacks the driving force for serving as a leader on this road. True, it is able to eliminate obstacles, but it cannot turn them into productive factors. For its activity is esentially critical, its glance is directed backward. For progress, advancement requires new associations of ideas and new queries, not based on the results of measurements alone, but going beyond them, and toward such things the fundamental atittude of Positivism is one of aloofness.

Therefore, up to quite recently, positivists have also put up the strongest resistance to the introduction of atomic hypotheses and thereby also to the acceptance of the above mentioned universal constants. This is quite understandable, for the existence of these constants is palpable proof of the existence in nature of something real and independent of every human measurement.

Of course, even today a consistent positivist could call the universal constants mere inventions which have proved to be uncommonly useful in making possible an accurate and complete description of the most diversified results of measurements. But hardly any real physicist would take such an assertion seriously. The universal constants were not invented for reasons of practical convenience, but have forced themselves upon us irresistibly because of the agreement between the results of all relevant measurements, and – this is the essential thing – we know quite well in advance that all future measurements will lead to these selfsame constants.

To sum it all up, we can say that physical science demands

that we admit the existence of a real world independent from us, a world which we can however never recognize directly but can apprehend only through the medium of our sense experiences and of the measurements mediated by them.

If we pursue this principle further, our outlook on the world takes a different form. The subject of the observation, the observing Ego, loses its position at the focus of thought and is relegated to a quite modest place. In fact, how pitifully small, how powerless we human beings must appear to ourselves if we stop to think that the planet Earth on which we live our lives is just a minute, infinitesimal mote of dust; on the other hand how peculiar it must seem that we, tiny creatures on a tiny planet, are nevertheless capable of knowing though not the essence at least the existence and the dimensions of the basic building blocks of the entire great Cosmos.

But this is still not the end of the wonder of it. Physical research has established as an incontestable fact that these basic building blocks of the Universe do not exist unrelated in isolated groups, but that all of them are mutually interlinked according to one uniform plan. In other words, every process in nature is subject to a universal and up to a point knowable law.

I want to mention at this place just one example: The law of the conservation of energy. There are various forms of energy in nature – kinetic energy, the energy of gravity, heat, electricity, magnetism. All the energies together form the energy supply of the world. The quantity of this energy supply is constant; it cannot be increased or diminished by any process in nature. All changes in nature are in reality simply the transformations of one form of energy into another. For instance, when kinetic energy is lost by friction, an equivalent quantity of thermal energy results.

The law of the conservation of energy is valid in every branch and field of physics, both according to the classical theory and to quantum mechanics. To be sure, there have often been attempts to challenge its precise validity in connection with the processes taking place within a single atom, and to assign to it a mere statistical significance for such processes. But every single experiment thus far undertaken has shown that such attempts are futile, and there is no reason to deny that the law of the conservation of energy is an absolutely and universally valid law of nature.

Those with positivistic leanings counter frequently with the critical objection that there is nothing extraordinary about the universal validity of such a law. According to them the explanation of the mystery is simply that after all it is man himself who prescribes the laws for nature. And in claiming this, they even cite Immanuel Kant in support of this view.

However, as I have pointed it out, the laws of nature were not invented by man, but external factors forced him to recognize them. An *a priori* approach to the laws of nature, as well as to the universal constants, could make us imagine them quite differently from what they are in reality. But the positivist reference to Kant is based on a gross misunderstanding. Kant did not teach that man actually prescribes laws for nature. He taught simply that whenever man formulates the laws of nature, he always adds something of his own, too. Otherwise how would it be conceivable that according to Kant's own statement, no external impression inspired in him a more profound feeling of respectful humility than the sight of the starry skies? After all, respectful humility is not exactly the attitude which a man is in the habit of assuming toward a rule formulated by himself. Obviously, such sentiment is alien to the mind of a positivist. To him the stars are nothing more than complexes of optical sensations; he considers everything else as just a useful but basically arbitrary and unessential trimming.

But let us now leave Positivism and continue with our own train of thought. The law of the conservation of energy is, after all, not the only law of nature, it is just one among many. While it is true that it is universally valid, it is still not sufficient for predicting every detail of a natural process, since it leaves an endless number of possibilities still open.

But there is another, far broader law, which has the property of giving a specific, unequivocal answer to each and every sensible question concerning the course of a natural process; as far as we can see, this law – like the law of the conservation of energy – possesses an exact validity even for the most modern parts of physics. But what we must regard as the gieatest wonder of all, is the fact that the most adequate formulation of this law creates the impression in every unbiased mind that nature is ruled by a rational, purposive will.

Let me illustrate this by a specific example. It is a well known fact that when a ray of light strikes the surface of a transparent body obliquely, such as the surface of water, it is deflected from its original direction after penetrating the surface. The explanation of this deflection is that light travels more slowly in water than in air. Such deflection, or "refraction," occurs also in the atmosphere, because light travels more slowly in the lower, denser strata of the asmosphere than in its higher layers. Now then, when a light ray emitted by a star reaches the eyes of an observer, its path will show a more or less complicated curve, due to the different degrees of refraction in the various atmospheric layers (unless the star happens to be exactly in the zenith). This curve is fully determined by the following simple law: Out of all the possible paths leading from the star to the eye of the observer, light will always follow the one which it can cover in the shortest time, allowance being made for the differences in its velocity in different atmospheric layers. Thus, the photons which constitute a ray of light behave like intelligent beings: Out of all the possible curves they always select the one which will take them most quickly to their goal.

This principle permits a large-scale generalization. According to all that we know about the laws relating to the processes taking place in any physical structure, we can characterize in all its details the course of each process by the principle that among all the conceivable processes which can change the state of a given physical structure into another state during a certain time interval, the process which actually takes place is always one for which the integjral over that time interval of a certain magnitude, the so-called Lagrange function, has the smallest value. Therefore, if we know the value of the Lagrange function, we can fully specify the course of the process actually taking place.

It is certainly no wonder that the discovery of this law – the so-called least-action principle, after which the elementary quantum of action was later also named – made its discoverer Leibniz, and soon after him also his follower Maupertuis, so boundlessly enthusiastic; for these scientists believed themselves to have found in it a tangible evidence for a ubiquitous higher reason ruling all nature.

In fact, the least-action principle introduces a completely new idea into the concept of causality: The *causa efficiens*, which operates from the present into the future and makes future situations appear as determined by earlier ones, is joined by the *causa finalis* for which, inversely, the future – namely, a definite goal – serves as the premise from which there can be deduced the development of the processes which lead to this goal.

So long as we confine ourselves to the realm of physics, these alternative points of view are merely different mathematical expressions for one and the same fact, and it would be futile to ask which of the two came nearer to the truth. The choice between them depends solely on practical considerations. The chief advantage of the least-action principle is that it requires no definite frame of reference for its formulation. This principle is therefore excellently adapted for carrying through transformation of coordinates.

But we are now interested in questions of a more general

character. It will suffice at this point to note only that the historical development of theoretic research in physics has led in a remarkable way to a formulation of the principle of physical causality which possesses an explicitly teleological character; but at the same time this formulation introduces nothing substantially new or even contradictory into the character of the laws of nature. The issue is simply one of different perspectives of interpretation, both of which are equally well justified by the actual facts. The situation in biology should be no different than we have found it in physics, although in biology the difference between the two viewpoints has assumed far sharper outlines.

In any case, we may say in summary that according to what exact natural science teaches us, the entire realm of nature, in which we human beings on our tiny mote of a planet play only an infinitesimally small part, is ruled by definite laws which are independent of the existence of thinking human beings; but these laws, insofar as they can at all be comprehended by our senses, can be given a formulation which is adapted for purposeful activity. Thus, natural science exhibits a rational world order to which nature and mankind are subject, but a world order the inner essence of which is and remains unknowable to us, since only our sense data (which can never be completely excluded) supply evidence for it. Nevertheless, the truly prolific results of natural scientific research justify the conclusion that continuing efforts will at least keep bringing us progressively nearer to the inattainable goal, and they strengthen our inner hope for a constant advancement of our insight into the ways of the omnipotent Reason which rules over Nature.

IV

Having now learned to know the demands which religion on one hand and science on the other hand place on our attitude to the most sublime problems of a generalized world outlook, let us now examine whether and to what extent these different demands can be mutually reconciled. First of all, it is self-evident that this examination may extend only to those laws in which religion and natural science conflict with each other. For these are wide spheres where they have absolutely nothing to do with each other. Thus, all the problems of ethics are outside of the field of natural science, whereas the dimensions of the universal constants are without relevance for religion.

On the other hand, religion and natural science do have a point of contact in the issue concerning the existence and nature of a supreme power ruling the world, and here the answers given by them are to a certain degree at least comparable. As we have seen, they are by no means mutually contradictory, but are in agreement, first of all, on the point that there exists a rational world order independent from man, and secondly, on the view that the character of this world order can never be directly known but can only be indirectly recognized or suspected. Religion employs in this connection its own characteristic symbols, while natural science uses measurements founded on sense experiences. Thus nothing stands in our way – and our instinctive intellectual striving for a unified world picture demands it – from identifying with each other the two everywhere active and yet mysterious forces: The world order of natural science and the God of religion. Accordingly, the deity which the religious person seeks to bring closer to, himself by his palpable symbols, is consubstantial with the power acting in accordance with natural laws for which the sense data of the scientist provide a certain degree of evidence.

However, in spite of this unanimity a fundamental difference must also be observed. To the religious person, God is directly and immediately given. He and His omnipotent Will are the fountainhead of all life and all happenings, both in the mundane world and in the world of the spirit. Even though

He cannot be grasped by reason, the religious symbols give a direct view of Him, and He plants His holy message in the souls of those who faithfully entrust themselves to Him. In contrast to this, the natural scientist recognizes as immediately given nothing but the content of his sense experiences and of the measurements based on them. He starts out from this point, on a road of inductive research, to approach as best he can the supreme and eternally unattainable goal of his quest – God and His world order. Therefore, while both religion and natural science require a belief in God for their activities, to the former He is the starting point, to the latter the goal of every thought process. To the former He is the foundation, to the latter the crown of the edifice of every generalized world view.

This difference corresponds to the different roles of religion and natural science in human life. Natural science wants man to learn, religion wants him to act. The only solid foundation for learning is the one supplied by sense perception; the assumption of a regular world order functions here merely as an essential condition for formulating fruitful questions. But this is not the road to be taken for action, for man's volitional decisions cannot wait until cognition has become complete or he has become omniscient. We stand in the midst of life, and its manifold demands and needs often make it imperative that we reach decisions or translate our mental attitudes into immediate action. Long and tedious reflection cannot enable us to shape our decisions and attitudes properly; only that definite and clear instruction can which we gain from a direct inner link to God. This instruction alone is able to give us the inner firmness and lasting peace of mind which must be regarded as the highest boon in life. And if we ascribe to God, in addition to His omnipotence and omniscience, also the attributes of goodness and love, recourse to Him produces an increased feeling of safety and happiness in the human being thirsting for solace. Against this conception not even the slightest objection can be raised from the point of view of natural science, for as we pointed it out before, questions of ethics are entirely outside of its realm.

No matter where and how far we look, nowhere do we find a contradiction between religion and natural science. On the contrary, we find a complete concordance in the very points of decisive importance. Religion and natural science do not exclude each other, as many contemporaries of ours would believe or fear; they mutually supplement and condition each other. The most immediate proof of the compatibility of religion and natural science, even under the most through critical scrutiny, is the historic fact that the very greatest natural scientists of all times – men such as Kepler, Newton, Leibniz – were permeated by a most profound religious attitude. At the dawn of our own era of civilization, the practitioners of natural science were the custodians of religion at the same time. The oldest of all the applied natural sciences, medicine, was in the hands of the priests, and in the Middle Ages scientific research was still carried on principally in monasteries. Later, as civilization continued to advance and to branch out, the parting of the ways became ahvays more pronounced, corresponding to the different nature of the tasks and pursuits of religion and those of natural science.

For the proper attitude to questions in ethics can no more be gained from a purely rational cognition than can a general *Weltanschauung* ever replace specific knowledge and ability. But the two roads do not diverge; they run parallel to each other, and they intersect at an endlessly removed common goal.

There is no better way to comprehend this properly than to continue one's efforts to obtain a progressively more profound insight into the nature and problems of the natural sciences, on one hand, and of religious faith on the other. It will then appear with ever increasing clarity that even though the methods are different – for science operates predominantly

146

with the intellect, religion predominantly with sentiment –
the significance of the work and the direction of progress are
nonetheless absolutely identical.

Religion and natural science are fighting a joint battle in
an incessant, never relaxing crusade against scepticism and
against dogmatism, against disbelief and against superstition,
and the rallying cry in this crusade has always been, and
always will be: *"On to God!"*

www.ingramcontent.com/pod-product-compliance
Lightning Source LLC
Chambersburg PA
CBHW021931190326
41519CB00009B/988